T0351179

Graduate Texts in Mathematics **58**

Springer
New York
Berlin
Heidelberg
Barcelona
Budapest
Hong Kong
London
Milan
Paris
Santa Clara
Singapore
Tokyo

Artist's conception of the 3-adic unit disk.

Drawing by A.T. Fomenko of Moscow State University, Moscow, U.S.S.R.

Neal Koblitz

p-adic Numbers, p-adic Analysis, and Zeta-Functions

Second Edition

 Springer

Neal Koblitz
Department of Mathematics GN-50
University of Washington
Seattle, WA 98195
USA

Mathematics Subject Classifications: 1991: 11-01, 11E95, 11Mxx

Library of Congress Cataloging in Publication Data
Koblitz, Neal, 1948–
    P-adic numbers, p-adic analysis and zeta-functions.
    (Graduate texts in mathematics; 58)
    Bibliography: p.
    Includes index.
    1. p-adic numbers. 2. p-adic analysis. 3. Functions,
Zeta I. Title. II. Series.
QA241.K674 1984   512'.74   84-5503

Printed in the United States of America.
Typeset by Composition House Ltd., Salisbury, England.
Printed and bound by Braun-Brumfield, Inc., Ann Arbor, MI.

9 8 7 6 5 4 3

ISBN 0-387-96017-1 Springer-Verlag New York Berlin Heidelberg
ISBN 3-540-96017-1 Springer-Verlag Berlin Heidelberg New York SPIN 10670823

To Professor Mark Kac

Preface to the second edition

The most important revisions in this edition are: (1) enlargement of the treatment of p-adic functions in Chapter IV to include the Iwasawa logarithm and the p-adic gamma-function, (2) rearrangement and addition of some exercises, (3) inclusion of an extensive appendix of answers and hints to the exercises, the absence of which from the first edition was apparently a source of considerable frustration for many readers, and (4) numerous corrections and clarifications, most of which were proposed by readers who took the trouble to write me. Some clarifications in Chapters IV and V were also suggested by V. V. Shokurov, the translator of the Russian edition. I am grateful to all of these readers for their assistance. I would especially like to thank Richard Bauer and Keith Conrad, who provided me with systematic lists of misprints and unclarities.

I would also like to express my gratitude to the staff of Springer-Verlag for both the high quality of their production and the cooperative spirit with which they have worked with me on this book and on other projects over the past several years.

Seattle, Washington   N. I. K.

Preface to the first edition

These lecture notes are intended as an introduction to p-adic analysis on the elementary level. For this reason they presuppose as little background as possible. Besides about three semesters of calculus, I presume some slight exposure to more abstract mathematics, to the extent that the student won't have an adverse reaction to matrices with entries in a field other than the real numbers, field extensions of the rational numbers, or the notion of a continuous map of topological spaces.

The purpose of this book is twofold: to develop some basic ideas of p-adic analysis, and to present two striking applications which, it is hoped, can be as effective pedagogically as they were historically in stimulating interest in the field. The first of these applications is presented in Chapter II, since it only requires the most elementary properties of $\mathbf{Q}_p$; this is Mazur's construction by means of p-adic integration of the Kubota–Leopoldt p-adic zeta-function, which "p-adically interpolates" the values of the Riemann zeta-function at the negative odd integers. My treatment is based on Mazur's Bourbaki notes (unpublished). The book then returns to the foundations of the subject, proving extension of the p-adic absolute value to algebraic extensions of $\mathbf{Q}_p$, constructing the p-adic analogue of the complex numbers, and developing the theory of p-adic power series. The treatment highlights analogies and contrasts with the familiar concepts and examples from calculus. The second main application, in Chapter V, is Dwork's proof of the rationality of the zeta-function of a system of equations over a finite field, one of the parts of the celebrated Weil Conjectures. Here the presentation follows Serre's exposition in *Séminaire Bourbaki*.

These notes have no pretension to being a thorough introduction to p-adic analysis. Such topics as the Hasse–Minkowski Theorem (which is in Chapter 1 of Borevich and Shafarevich's *Number Theory*) and Tate's thesis (which is also available in textbook form, see Lang's *Algebraic Number Theory*) are omitted.

Moreover, there is no attempt to present results in their most general form. For example, p-adic L-functions corresponding to Dirichlet characters are only discussed parenthetically in Chapter II. The aim is to present a selection of material that can be digested by undergraduates or beginning graduate students in a one-term course.

The exercises are for the most part not hard, and are important in order to convert a passive understanding to a real grasp of the material. The abundance of exercises will enable many students to study the subject on their own, with minimal guidance, testing themselves and solidifying their understanding by working the problems.

p-adic analysis can be of interest to students for several reasons. First of all, in many areas of mathematical research—such as number theory and representation theory—p-adic techniques occupy an important place. More naively, for a student who has just learned calculus, the "brave new world" of non-Archimedean analysis provides an amusing perspective on the world of classical analysis. p-adic analysis, with a foot in classical analysis and a foot in algebra and number theory, provides a valuable point of view for a student interested in any of those areas.

I would like to thank Professors Mark Kac and Yu. I. Manin for their help and encouragement over the years, and for providing, through their teaching and writing, models of pedagogical insight which their students can try to emulate.

Logical dependence of chapters

Cambridge, Massachusetts   N. I. K.

Contents

Contents

CHAPTER I

p-adic numbers

1. Basic concepts

If X is a nonempty set, a distance, or *metric*, on X is a function d from pairs of elements (x, y) of X to the nonnegative real numbers such that

(1) $d(x, y) = 0$ if and only if $x = y$.
(2) $d(x, y) = d(y, x)$.
(3) $d(x, y) \le d(x, z) + d(z, y)$ for all $z \in X$.

A set X together with a metric d is called a *metric space*. The same set X can give rise to many different metric spaces (X, d), as we'll soon see.

The sets X we'll be dealing with will mostly be fields. Recall that a field F is a set together with two operations $+$ and $\cdot$ such that F is a commutative group under $+$, $F - \{0\}$ is a commutative group under $\cdot$, and the distributive law holds. The examples of a field to have in mind at this point are the field $\mathbb{Q}$ of rational numbers and the field $\mathbb{R}$ of real numbers.

The metrics d we'll be dealing with will come from *norms* on the field F, which means a map denoted $\| \ \|$ from F to the nonnegative real numbers such that

(1) $\|x\| = 0$ if and only if $x = 0$.
(2) $\|x \cdot y\| = \|x\| \cdot \|y\|$.
(3) $\|x + y\| \le \|x\| + \|y\|$.

When we say that a metric d "comes from" (or "is induced by") a norm $\| \ \|$, we mean that d is defined by: $d(x, y) = \|x - y\|$. It is an easy exercise to check that such a d satisfies the definition of a metric whenever $\| \ \|$ is a norm.

A basic example of a norm on the rational number field $\mathbb{Q}$ is the absolute value $|x|$. The induced metric $d(x, y) = |x - y|$ is the usual concept of distance on the number line.

1

My reason for starting with the abstract definition of distance is that the point of departure for our whole subject of study will be a new type of distance, which will satisfy Properties (1)–(3) in the definition of a metric but will differ fundamentally from the familiar intuitive notions. My reason for recalling the abstract definition of a field is that we'll soon need to be working not only with **Q** but with various "extension fields" which contain **Q**.

2. Metrics on the rational numbers

We know one metric on **Q**, that induced by the ordinary absolute value. Are there any others? The following is basic to everything that follows.

Definition. Let $p \in \{2, 3, 5, 7, 11, 13, \ldots\}$ be any prime number. For any nonzero integer a, let the *p*-adic ordinal of a, denoted $\mathrm{ord}_p\, a$, be the highest power of p which divides a, i.e., the greatest m such that $a \equiv 0 \pmod{p^m}$. (The notation $a \equiv b \pmod c$ means: c divides $a - b$.) For example,

$$\mathrm{ord}_5 35 = 1, \qquad \mathrm{ord}_5 250 = 3, \qquad \mathrm{ord}_2 96 = 5, \qquad \mathrm{ord}_2 97 = 0.$$

(If $a = 0$, we agree to write $\mathrm{ord}_p 0 = \infty$.) Note that ord_p behaves a little like a logarithm would: $\mathrm{ord}_p(a_1 a_2) = \mathrm{ord}_p\, a_1 + \mathrm{ord}_p\, a_2$.

Now for any rational number $x = a/b$, define $\mathrm{ord}_p\, x$ to be $\mathrm{ord}_p\, a - \mathrm{ord}_p\, b$. Note that this expression depends only on x, and not on a and b, i.e., if we write $x = ac/bc$, we get the same value for $\mathrm{ord}_p\, x = \mathrm{ord}_p\, ac - \mathrm{ord}_p\, bc$.

Further define a map $|\ |_p$ on **Q** as follows:

$$|x|_p = \begin{cases} \dfrac{1}{p^{\mathrm{ord}_p x}}, & \text{if } x \neq 0; \\ 0, & \text{if } x = 0. \end{cases}$$

Proposition. $|\ |_p$ *is a norm on* **Q**.

PROOF. Properties (1) and (2) are easy to check as an exercise. We now verify (3).

If $x = 0$ or $y = 0$, or if $x + y = 0$, Property (3) is trivial, so assume x, y, and $x + y$ are all nonzero. Let $x = a/b$ and $y = c/d$ be written in lowest terms. Then we have: $x + y = (ad + bc)/bd$, and $\mathrm{ord}_p(x + y) = \mathrm{ord}_p(ad + bc) - \mathrm{ord}_p\, b - \mathrm{ord}_p\, d$. Now the highest power of p dividing the sum of two numbers is *at least* the minimum of the highest power dividing the first and the highest power dividing the second. Hence

$$\begin{aligned}
\mathrm{ord}_p(x + y) &\geq \min(\mathrm{ord}_p\, ad, \mathrm{ord}_p\, bc) - \mathrm{ord}_p\, b - \mathrm{ord}_p\, d \\
&= \min(\mathrm{ord}_p\, a + \mathrm{ord}_p\, d, \mathrm{ord}_p\, b + \mathrm{ord}_p\, c) - \mathrm{ord}_p\, b - \mathrm{ord}_p\, d \\
&= \min(\mathrm{ord}_p\, a - \mathrm{ord}_p\, b, \mathrm{ord}_p\, c - \mathrm{ord}_p\, d) \\
&= \min(\mathrm{ord}_p\, x, \mathrm{ord}_p\, y).
\end{aligned}$$

Therefore, $|x + y|_p = p^{-\mathrm{ord}_p(x+y)} \leq \max(p^{-\mathrm{ord}_p x}, p^{-\mathrm{ord}_p y}) = \max(|x|_p, |y|_p)$, and this is $\leq |x|_p + |y|_p$. □

We actually proved a stronger inequality than Property (3), and it is this stronger inequality which leads to *the* basic definition of *p*-adic analysis.

Definition. A norm is called *non-Archimedean* if $\|x + y\| \leq \max(\|x\|, \|y\|)$ always holds. A metric is called *non-Archimedean* if $d(x, y) \leq \max(d(x, z), d(z, y))$; in particular, a metric is non-Archimedean if it is induced by a non-Archimedean norm, since in that case $d(x, y) = \|x - y\| = \|(x - z) + (z - y)\| \leq \max(\|x - z\|, \|z - y\|) = \max(d(x, z), d(z, y))$.

Thus, $|\ |_p$ is a non-Archimedean norm on $\mathbb{Q}$.

A norm (or metric) which is not non-Archimedean is called *Archimedean*. The ordinary absolute value is an Archimedean norm on $\mathbb{Q}$.

In any metric space X we have the notion of a *Cauchy sequence* $\{a_1, a_2, a_3, \ldots\}$ of elements of X. This means that for any $\varepsilon > 0$ there exists an N such that $d(a_m, a_n) < \varepsilon$ whenever both $m > N$ and $n > N$.

We say two metrics d_1 and d_2 on a set X are *equivalent* if a sequence is Cauchy with respect to d_1 if and only if it is Cauchy with respect to d_2. We say two norms are *equivalent* if they induce equivalent metrics.

In the definition of $|\ |_p$, instead of $(1/p)^{\mathrm{ord}_p x}$ we could have written $\rho^{\mathrm{ord}_p x}$ with any $\rho \in (0, 1)$ in place of $1/p$. We would have obtained an equivalent non-Archimedean norm (see Exercises 5 and 6). The reason why $\rho = 1/p$ is usually the most convenient choice is related to the formula in Exercise 18 below.

We also have a family of Archimedean norms which are equivalent to the usual absolute value $|\ |$, namely $|\ |^\alpha$ when $0 < \alpha \leq 1$ (see Exercise 8).

We sometimes let $|\ |_\infty$ denote the usual absolute value. This is only a notational convention, and is not meant to imply any direct relationship between $|\ |_\infty$ and $|\ |_p$.

By the "trivial" norm we mean the norm $\|\ \|$ such that $\|0\| = 0$ and $\|x\| = 1$ for $x \neq 0$.

Theorem 1 (Ostrowski). *Every nontrivial norm $\|\ \|$ on $\mathbb{Q}$ is equivalent to $|\ |_p$ for some prime p or for $p = \infty$.*

PROOF. *Case* (i). Suppose there exists a positive integer n such that $\|n\| > 1$. Let n_0 be the least such n. Since $\|n_0\| > 1$, there exists a positive real number α such that $\|n_0\| = n_0^\alpha$. Now write any positive integer n to the base n_0, i.e., in the form

$$n = a_0 + a_1 n_0 + a_2 n_0^2 + \cdots + a_s n_0^s, \quad \text{where } 0 \leq a_i < n_0 \text{ and } a_s \neq 0.$$

Then

$$\|n\| \leq \|a_0\| + \|a_1 n_0\| + \|a_2 n_0^2\| + \cdots + \|a_s n_0^s\|$$
$$= \|a_0\| + \|a_1\| \cdot n_0^\alpha + \|a_2\| \cdot n_0^{2\alpha} + \cdots + \|a_s\| \cdot n_0^{s\alpha}.$$

Since all of the a_i are $< n_0$, by our choice of n_0 we have $\|a_i\| \leq 1$, and hence

$$\|n\| \leq 1 + n_0^\alpha + n_0^{2\alpha} + \cdots + n_0^{s\alpha}$$
$$= n_0^{s\alpha}(1 + n_0^{-\alpha} + n_0^{-2\alpha} + \cdots + n_0^{-s\alpha})$$
$$\leq n^\alpha \left[\sum_{i=0}^\infty (1/n_0^\alpha)^i \right],$$

because $n \geq n_0^s$. The expression in brackets is a finite constant, which we call C. Thus,

$$\|n\| \leq Cn^\alpha \quad \text{for all } n = 1, 2, 3, \ldots.$$

Now take any n and any large N, and put n^N in place of n in the above inequality; then take Nth roots. You get

$$\|n\| \leq \sqrt[N]{C} n^\alpha.$$

Letting $N \to \infty$ for n fixed gives $\|n\| \leq n^\alpha$.

We can get the inequality the other way as follows. If n is written to the base n_0 as before, we have $n_0^{s+1} > n \geq n_0^s$. Since $\|n_0^{s+1}\| = \|n + n_0^{s+1} - n\| \leq \|n\| + \|n_0^{s+1} - n\|$, we have

$$\|n\| \geq \|n_0^{s+1}\| - \|n_0^{s+1} - n\|$$
$$\geq n_0^{(s+1)\alpha} - (n_0^{s+1} - n)^\alpha,$$

since $\|n_0^{s+1}\| = \|n_0\|^{s+1}$, and we can use the first inequality (i.e., $\|n\| \leq n^\alpha$) on the term that is being subtracted. Thus,

$$\|n\| \geq n_0^{(s+1)\alpha} - (n_0^{s+1} - n_0^s)^\alpha \quad (\text{since } n \geq n_0^s)$$
$$= n_0^{(s+1)\alpha} \left[1 - \left(1 - \frac{1}{n_0} \right)^\alpha \right]$$
$$\geq C'n^\alpha$$

for some constant C' which may depend on n_0 and α but not on n. As before, we now use this inequality for n^N, take Nth roots, and let $N \to \infty$, finally getting: $\|n\| \geq n^\alpha$.

Thus, $\|n\| = n^\alpha$. It easily follows from Property (2) of norms that $\|x\| = |x|^\alpha$ for all $x \in \mathbb{Q}$. In view of Exercise 8 below, which says that such a norm is equivalent to the absolute value $|\ |$, this concludes the proof of the theorem in Case (i).

Case (ii). Suppose that $\|n\| \leq 1$ for all positive integers n. Let n_0 be the least n such that $\|n\| < 1$; n_0 exists because we have assumed that $\|\ \|$ is nontrivial.

n_0 must be a prime, because if $n_0 = n_1 \cdot n_2$ with n_1 and n_2 both $< n_0$, then $\|n_1\| = \|n_2\| = 1$, and so $\|n_0\| = \|n_1\| \cdot \|n_2\| = 1$. So let p denote the prime n_0.

We claim that $\|q\| = 1$ if q is a prime not equal to p. Suppose not; then $\|q\| < 1$, and for some large N we have $\|q^N\| = \|q\|^N < \frac{1}{2}$. Also, for some large M we have $\|p^M\| < \frac{1}{2}$. Since p^M and q^N are relatively prime—have no common divisor other than 1—we can find (see Exercise 10) integers n and m such that: $mp^M + nq^N = 1$. But then

$$1 = \|1\| = \|mp^M + nq^N\| \le \|mp^M\| + \|nq^N\| = \|m\| \, \|p^M\| + \|n\| \, \|q^N\|,$$

by Properties (2) and (3) in the definition of a norm. But $\|m\|$, $\|n\| \le 1$, so that

$$1 \le \|p^M\| + \|q^N\| < \tfrac{1}{2} + \tfrac{1}{2} = 1,$$

a contradiction. Hence $\|q\| = 1$.

We're now virtually done, since any positive integer a can be factored into prime divisors: $a = p_1^{b_1} p_2^{b_2} \cdots p_r^{b_r}$. Then $\|a\| = \|p_1\|^{b_1} \cdot \|p_2\|^{b_2} \cdots \|p_r\|^{b_r}$. But the only $\|p_i\|$ which is not equal to 1 will be $\|p\|$ if one of the p_i's is p. Its corresponding b_i will be $\mathrm{ord}_p\, a$. Hence, if we let $\rho = \|p\| < 1$, we have

$$\|a\| = \rho^{\mathrm{ord}_p a}.$$

It is easy to see using Property (2) of a norm that the same formula holds with any nonzero rational number x in place of a. In view of Exercise 5 below, which says that such a norm is equivalent to $|\;|_p$, this concludes the proof of Ostrowski's theorem. □

Our intuition about distance is based, of course, on the Archimedean metric $|\;|_\infty$. Some properties of the non-Archimedean metrics $|\;|_p$ seem very strange at first, and take a while to get used to. Here are two examples.

For any metric, Property (3): $d(x, y) \le d(x, z) + d(z, y)$ is known as the "triangle inequality," because in the case of the field $\mathbb{C}$ of complex numbers (with metric $d(a + bi, c + di) = \sqrt{(a - c)^2 + (b - d)^2}$) it says that in the complex plane the sum of two sides of a triangle is greater than the third side. (See the diagram.)

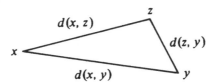

Let's see what happens with a non-Archimedean norm on a field F. For simplicity suppose $z = 0$. Then the non-Archimedean triangle inequality says: $\|x - y\| \le \max(\|x\|, \|y\|)$. Suppose first that the "sides" x and y have different "length," say $\|x\| < \|y\|$. The third side $x - y$ has length

$$\|x - y\| \le \|y\|.$$

But

$$\|y\| = \|x - (x - y)\| \le \max(\|x\|, \|x - y\|).$$

Since $\|y\|$ is not $\le \|x\|$, we must have $\|y\| \le \|x - y\|$, and so $\|y\| = \|x - y\|$.

Thus, if our two sides x and y are not equal in length, the longer of the two must have the same length as the third side. Every "triangle" is isosceles!

This really shouldn't be too surprising if we think what this says in the case of $| \ |_p$ on $\mathbf{Q}$. It says that, if two rational numbers are divisible by different powers of p, then their difference is divisible precisely by the *lower* power of p (which is what it means to be the same "size" as the *bigger* of the two).

This basic property of a non-Archimedean field—that $\|x \pm y\| \le \max(\|x\|, \|y\|)$, with equality holding if $\|x\| \ne \|y\|$—will be referred to as the "isosceles triangle principle" from now on.

As a second example, we define the (open) disc of radius r (r is a positive real number) with center a (a is an element in the field F) to be

$$D(a, r^-) = \{x \in F \mid \|x - a\| < r\}.$$

Suppose $\| \ \|$ is a non-Archimedean norm. Let b be *any* element in $D(a, r^-)$. Then

$$D(a, r^-) = D(b, r^-),$$

i.e., every point in the disc is a center! Why is this? Well

$$
\begin{aligned}
x \in D(a, r^-) &\Rightarrow \|x - a\| < r \\
&\Rightarrow \|x - b\| = \|(x - a) + (a - b)\| \\
&\qquad\qquad \le \max(\|x - a\|, \|a - b\|) \\
&\qquad\qquad < r \\
&\Rightarrow x \in D(b, r^-),
\end{aligned}
$$

and the reverse implication is proved in the exact same way.

If we define the closed disc of radius r with center a to be

$$D(a, r) = \{x \in F \mid \|x - a\| \le r\},$$

for non-Archimedean $\| \ \|$ we similarly find that every point in $D(a, r)$ is a center.

EXERCISES

1. For any norm $\| \ \|$ on a field F, prove that addition, multiplication, and finding the additive and multiplicative inverses are continuous. This means that: (1) for any $x, y \in F$ and any $\varepsilon > 0$, there exists $\delta > 0$ such that $\|x' - x\| < \delta$ and $\|y' - y\| < \delta$ imply $\|(x' + y') - (x + y)\| < \varepsilon$; (2) the same statement with $\|(x' + y') - (x + y)\|$ replaced by $\|x'y' - xy\|$; (3) for any nonzero $x \in F$ and any $\varepsilon > 0$, there exists $\delta > 0$ such that $\|x' - x\| < \delta$ implies $\|(1/x') - (1/x)\| < \varepsilon$; (4) for any $x \in F$ and any $\varepsilon > 0$, there exists $\delta > 0$ such that $\|x' - x\| < \delta$ implies $\|(-x') - (-x)\| < \varepsilon$.

2. Prove that if $\| \ \|$ is any norm on a field F, then $\|-1\| = \|1\| = 1$. Prove that if $\| \ \|$ is non-Archimedean, then for any integer n: $\|n\| \le 1$. (Here "n" means the result of adding $1 + 1 + 1 + \cdots + 1$ together n times in the field F.)

3. Prove that, conversely, if $\| \ \|$ is a norm such that $\|n\| \leq 1$ for every integer n, then $\| \ \|$ is non-Archimedean.

4. Prove that a norm $\| \ \|$ on a field F is non-Archimedean if and only if

$$\{x \in F \,|\, \|x\| < 1\} \cap \{x \in F \,|\, \|x - 1\| < 1\} = \varnothing.$$

5. Let $\| \ \|_1$ and $\| \ \|_2$ be two norms on a field F. Prove that $\| \ \|_1 \sim \| \ \|_2$ if and only if there exists a positive real number α such that: $\|x\|_1 = \|x\|_2{}^\alpha$ for all $x \in F$.

6. Prove that, if $0 < \rho < 1$, then the function on $x \in \mathbb{Q}$ defined as $\rho^{\text{ord}_p x}$ if $x \neq 0$ and 0 if $x = 0$, is a non-Archimedean norm. Note that by the previous problem it is equivalent to $| \ |_p$. What happens if $\rho = 1$? What about if $\rho > 1$?

7. Prove that $| \ |_{p_1}$ is not equivalent to $| \ |_{p_2}$ if p_1 and p_2 are different primes.

8. For $x \in \mathbb{Q}$ define $\|x\| = |x|^\alpha$ for a fixed positive number α, where $| \ |$ is the usual absolute value. Show that $\| \ \|$ is a norm if and only if $\alpha \leq 1$, and that in that case it is equivalent to the norm $| \ |$.

9. Prove that two equivalent norms on a field F are either both non-Archimedean or both Archimedean.

10. Prove that, if N and M are relatively prime integers, then there exist integers n and m such that $nN + mM = 1$.

11. Evaluate:

(i) $\text{ord}_3 \, 54$	(ii) $\text{ord}_2 \, 128$	(iii) $\text{ord}_3 \, 57$
(iv) $\text{ord}_7(-700/197)$	(v) $\text{ord}_2(128/7)$	(vi) $\text{ord}_3(7/9)$
(vii) $\text{ord}_5(-0.0625)$	(viii) $\text{ord}_3(10^9)$	(ix) $\text{ord}_3(-13.23)$
(x) $\text{ord}_7(-13.23)$	(xi) $\text{ord}_5(-13.23)$	(xii) $\text{ord}_{11}(-13.23)$
(xiii) $\text{ord}_{13}(-26/169)$	(xiv) $\text{ord}_{103}(-1/309)$	(xv) $\text{ord}_3(9!)$

12. Prove that $\text{ord}_p((p^N)!) = 1 + p + p^2 + \cdots + p^{N-1}$.

13. If $0 \leq a \leq p - 1$, prove that: $\text{ord}_p((ap^N)!) = a(1 + p + p^2 + \cdots + p^{N-1})$.

14. Prove that, if $n = a_0 + a_1 p + a_2 p^2 + \cdots + a_s p^s$ is written to the base p, so that $0 \leq a_i \leq p - 1$, and if we set $S_n = \sum a_i$ (the sum of the digits to the base p), then we have the formula:

$$\text{ord}_p(n!) = \frac{n - S_n}{p - 1}.$$

15. Evaluate $|a - b|_p$, i.e., the p-adic distance between a and b, when:

(i) $a = 1, b = 26, p = 5$	(ii) $a = 1, b = 26, p = \infty$
(iii) $a = 1, b = 26, p = 3$	(iv) $a = 1/9, b = -1/16, p = 5$
(v) $a = 1, b = 244, p = 3$	(vi) $a = 1, b = 1/244, p = 3$
(vii) $a = 1, b = 1/243, p = 3$	(viii) $a = 1, b = 183, p = 13$
(ix) $a = 1, b = 183, p = 7$	(x) $a = 1, b = 183, p = 2$
(xi) $a = 1, b = 183, p = \infty$	(xii) $a = 9!, b = 0, p = 3$
(xiii) $a = (9!)^2/3^9, b = 0, p = 3$	(xiv) $a = 2^{2N}/2^N, b = 0, p = 2$
(xv) $a = 2^{2N}/(2^N)!, b = 0, p = 2.$	

16. Say in words what it means for a rational number x to satisfy $|x|_p \leq 1$.

17. For $x \in \mathbb{Q}$, prove that $\lim_{i \to \infty} |x^i/i!|_p = 0$ if and only if: $\text{ord}_p x \geq 1$ when $p \neq 2$, $\text{ord}_2 x \geq 2$ when $p = 2$.

18. Let x be a nonzero rational number. Prove that the product over *all* primes *including* ∞ of $|x|_p$ equals 1. (Notice that this "infinite product" actually only includes a finite number of terms that are not equal to 1.) Symbolically, $\prod_p |x|_p = 1$.

19. Prove that for any p ($\neq \infty$), any sequence of integers has a subsequence which is Cauchy with respect to $| \ |_p$.

20. Prove that if $x \in \mathbb{Q}$ and $|x|_p \leq 1$ for every prime p, then $x \in \mathbb{Z}$.

3. Review of building up the complex numbers

We now have a new concept of distance between two rational numbers: two rational numbers are considered to be close if their difference is divisible by a large power of a fixed prime p. In order to work with this so-called "p-adic metric" we must enlarge the rational number field $\mathbb{Q}$ in a way analogous to how the real numbers $\mathbb{R}$ and then the complex numbers $\mathbb{C}$ were constructed in the classical Archimedean metric $| \ |$. So let's review how this was done.

Let's go back even farther, logically and historically, than $\mathbb{Q}$. Let's go back to the natural numbers $\mathbb{N} = \{1, 2, 3, \ldots\}$. Every step in going from $\mathbb{N}$ to $\mathbb{C}$ can be analyzed in terms of a desire to do two things:

(1) Solve polynomial equations.
(2) Find limits of Cauchy sequences, i.e., "complete" the number system to one "without holes," in which every Cauchy sequence has a limit in the new number system.

First of all, the integers $\mathbb{Z}$ (including $0, -1, -2, \ldots$) can be introduced as solutions of equations of the form

$$a + x = b, \quad a, b \in \mathbb{N}.$$

Next, rational numbers can be introduced as solutions of equations of the form

$$ax = b, \quad a, b \in \mathbb{Z}.$$

So far we haven't used any concept of distance.

One of the possible ways to give a careful definition of the real numbers is to consider the set S of Cauchy sequences of rational numbers. Call two Cauchy sequences $s_1 = \{a_j\} \in S$ and $s_2 = \{b_j\} \in S$ equivalent, and write $s_1 \sim s_2$, if $|a_j - b_j| \to 0$ as $j \to \infty$. This is obviously an equivalence relation, that is, we have: (1) any s is equivalent to itself; (2) if $s_1 \sim s_2$, then $s_2 \sim s_1$; and (3) if $s_1 \sim s_2$ and $s_2 \sim s_3$, then $s_1 \sim s_3$. We then define $\mathbb{R}$ to be the set of *equivalence classes* of Cauchy sequences of rational numbers. It is not hard to define addition, multiplication, and finding additive and multiplicative inverses of equivalence classes of Cauchy sequences, and to show that $\mathbb{R}$ is a field. Even though this definition seems rather abstract and cumbersome at first glance, it turns out that it gives no more nor less than the old-fashioned real number line, which is so easy to visualize.

Something similar will happen when we work with $|\ |_p$ instead of $|\ |$: starting with an abstract definition of the p-adic completion of $\mathbb{Q}$, we'll get a very down-to-earth number system, which we'll call $\mathbb{Q}_p$.

Getting back to our historical survey, we've gotten as far as $\mathbb{R}$. Next, returning to the first method—solving equations—mathematicians decided that it would be a good idea to have numbers that could solve equations like $x^2 + 1 = 0$. (This is taking things in logical order; historically speaking, the definition of the complex numbers came before the rigorous definition of the real numbers in terms of Cauchy sequences.) Then an amazing thing happened! As soon as $i = \sqrt{-1}$ was introduced and the field of complex numbers of the form $a + bi$, $a, b \in \mathbb{R}$, was defined, it turned out that:

(1) *All* polynomial equations with coefficients in $\mathbb{C}$ have solutions in $\mathbb{C}$—this is the famous Fundamental Theorem of Algebra (the concise terminology is to say that $\mathbb{C}$ is *algebraically closed*); and
(2) $\mathbb{C}$ is already "complete" with respect to the (unique) norm which extends the norm $|\ |$ on $\mathbb{R}$ (this norm is given by $|a + bi| = \sqrt{a^2 + b^2}$), i.e., any Cauchy sequence $\{a_j + b_j i\}$ has a limit of the form $a + bi$ (since $\{a_j\}$ and $\{b_j\}$ will each be Cauchy sequences in $\mathbb{R}$, you just let a and b be their limits).

So the process stops with $\mathbb{C}$, which is only a "quadratic extension" of $\mathbb{R}$ (i.e., obtained by adjoining a solution of the quadratic equation $x^2 + 1 = 0$). $\mathbb{C}$ *is an algebraically closed field which is complete with respect to the Archimedean metric.*

But alas! Such is not to be the case with $|\ |_p$. After getting $\mathbb{Q}_p$, the completion of $\mathbb{Q}$ with respect to $|\ |_p$, we must then form an *infinite* sequence of field extensions obtained by adjoining solutions to higher degree (not just quadratic) equations. Even worse, the resulting algebraically closed field, which we denote $\overline{\mathbb{Q}}_p$, is *not complete*. So we take this already gigantic field and "fill in the holes" to get a still larger field Ω.

What happens then? Do we now have to enlarge Ω to be able to solve polynomial equations with coefficients in Ω? Does this process continue on and on, in a frightening spiral of ever more far-fetched abstractions? Well, fortunately, with Ω the guardian angel of p-adic analysis intervenes, and it turns out that Ω is already algebraically closed, as well as complete, and our search for the non-Archimedean analogue of $\mathbb{C}$ is ended.

But this Ω, which will be the convenient number system in which to study the p-adic analogy of calculus and analysis, is much less thoroughly understood than $\mathbb{C}$. As I. M. Gel'fand has remarked, some of the simplest questions, e.g., characterizing $\mathbb{Q}_p$-linear field automorphisms of Ω, remain unanswered.

So let's begin our journey to Ω.

4. The field of *p*-adic numbers

For the rest of this chapter, we fix a prime number $p \neq \infty$.

Let S be the set of sequences $\{a_i\}$ of rational numbers such that, given $\varepsilon > 0$, there exists an N such that $|a_i - a_{i'}|_p < \varepsilon$ if both $i, i' > N$. We call two such Cauchy sequences $\{a_i\}$ and $\{b_i\}$ equivalent if $|a_i - b_i|_p \to 0$ as $i \to \infty$. We define the set $\mathbb{Q}_p$ to be the set of equivalence classes of Cauchy sequences.

For any $x \in \mathbb{Q}$, let $\{x\}$ denote the "constant" Cauchy sequence all of whose terms equal x. It is obvious that $\{x\} \sim \{x'\}$ if and only if $x = x'$. The equivalence class of $\{0\}$ is denoted simply by 0.

We define the norm $|\ |_p$ of an equivalence class a to be $\lim_{i \to \infty} |a_i|_p$, where $\{a_i\}$ is any representative of a. The limit exists because

(1) If $a = 0$, then by definition $\lim_{i \to \infty} |a_i|_p = 0$.
(2) If $a \neq 0$, then for some ε and for every N there exists an $i_N > N$ with $|a_{i_N}|_p > \varepsilon$.

If we choose N large enough so that $|a_i - a_{i'}|_p < \varepsilon$ when $i, i' > N$, we have:

$$|a_i - a_{i_N}|_p < \varepsilon \quad \text{for all } i > N.$$

Since $|a_{i_N}|_p > \varepsilon$, it follows by the "isosceles triangle principle" that $|a_i|_p = |a_{i_N}|_p$. Thus, for all $i > N$, $|a_i|_p$ has the constant value $|a_{i_N}|_p$. This constant value is then $\lim_{i \to \infty} |a_i|_p$.

One important difference with the process of completing $\mathbb{Q}$ to get $\mathbb{R}$ should be noted. In going from $\mathbb{Q}$ to $\mathbb{R}$ the possible values of $|\ | = |\ |_\infty$ were enlarged to include all nonnegative real numbers. But in going from $\mathbb{Q}$ to $\mathbb{Q}_p$ the possible values of $|\ |_p$ remain the same, namely $\{p^n\}_{n \in \mathbb{Z}} \cup \{0\}$.

Given two equivalence classes a and b of Cauchy sequences, we choose any representatives $\{a_i\} \in a$ and $\{b_i\} \in b$, and define $a \cdot b$ to be the equivalence class represented by the Cauchy sequence $\{a_i b_i\}$. If we had chosen another $\{a_i'\} \in a$ and $\{b_i'\} \in b$, we would have

$$|a_i' b_i' - a_i b_i|_p = |a_i'(b_i' - b_i) + b_i(a_i' - a_i)|_p$$
$$\leq \max(|a_i'(b_i' - b_i)|_p, |b_i(a_i' - a_i)|_p);$$

as $i \to \infty$, the first expression approaches $|a|_p \cdot \lim |b_i' - b_i|_p = 0$, and the second expression approaches $|b|_p \cdot \lim |a_i' - a_i|_p = 0$. Hence $\{a_i' b_i'\} \sim \{a_i b_i\}$.

We similarly define the sum of two equivalence classes of Cauchy sequences by choosing a Cauchy sequence in each class, defining addition term-by-term, and showing that the equivalence class of the sum only depends on the equivalence classes of the two summands. Additive inverses are also defined in the obvious way.

For multiplicative inverses we have to be a little careful because of the possibility of zero terms in a Cauchy sequence. However, it is easy to see that every Cauchy sequence is equivalent to one with no zero terms (for example, if $a_i = 0$, replace a_i by $a_i' = p^i$). Then take the sequence $\{1/a_i\}$. This sequence will be Cauchy *unless* $|a_i|_p \to 0$, i.e., unless $\{a_i\} \sim \{0\}$. Moreover, if $\{a_i\} \sim \{a_i'\}$ and no a_i or a_i' is zero, then $\{1/a_i\} \sim \{1/a_i'\}$ is easily proved.

It is now easy to prove that the set $\mathbb{Q}_p$ of equivalence classes of Cauchy sequences is a field with addition, multiplication, and inverses defined as above. For example, distributivity: Let $\{a_i\}$, $\{b_i\}$, $\{c_i\}$ be representatives of $a, b, c \in \mathbb{Q}_p$; then $a(b + c)$ is the equivalence class of

$$\{a_i(b_i + c_i)\} = \{a_ib_i + a_ic_i\},$$

and $ab + ac$ is also the equivalence class of this sequence.

$\mathbb{Q}$ can be identified with the *subfield* of $\mathbb{Q}_p$ consisting of equivalence classes containing a constant Cauchy sequence. Under this identification, note that $|\ |_p$ on $\mathbb{Q}_p$ restricts to the usual $|\ |_p$ on $\mathbb{Q}$.

Finally, it is easy to prove that $\mathbb{Q}_p$ is complete: if $\{a_j\}_{j=1,2,\ldots}$ is a sequence of equivalence classes which is Cauchy in $\mathbb{Q}_p$, and if we take representative Cauchy sequences of rational numbers $\{a_{ji}\}_{i=1,2,\ldots}$ for each a_j, where for each j we have $|a_{ji} - a_{ji'}|_p < p^{-j}$ whenever $i, i' \geq N_j$, then it is easily shown that the equivalence class of $\{a_{jN_j}\}_{j=1,2,\ldots}$ is the limit of the a_j. We leave the details to the reader.

It's probably a good idea to go through one such tedious construction in any course or seminar, so as not to totally forget the axiomatic foundations on which everything rests. In this particular case, the abstract approach also gives us the chance to compare the p-adic construction with the construction of the reals, and see that the procedure is logically the same. However, after the following theorem, it would be wise to forget as rapidly as possible about "equivalence classes of Cauchy sequences," and to start thinking in more concrete terms.

Theorem 2. *Every equivalence class a in $\mathbb{Q}_p$ for which $|a|_p \leq 1$ has exactly one representative Cauchy sequence of the form $\{a_i\}$ for which:*

(1) $0 \leq a_i < p^i$ *for* $i = 1, 2, 3, \ldots$.
(2) $a_i \equiv a_{i+1} \pmod{p^i}$ *for* $i = 1, 2, 3, \ldots$.

PROOF. We first prove uniqueness. If $\{a_i'\}$ is a different sequence satisfying (1) and (2), and if $a_{i_0} \neq a_{i_0}'$, then $a_{i_0} \not\equiv a_{i_0}' \pmod{p^{i_0}}$, because both are between 0 and p^{i_0}. But then, for all $i \geq i_0$, we have $a_i \equiv a_{i_0} \not\equiv a_{i_0}' \equiv a_i' \pmod{p^{i_0}}$, i.e., $a_i \not\equiv a_i' \pmod{p^{i_0}}$. Thus

$$|a_i - a_i'|_p > 1/p^{i_0}$$

for all $i \geq i_0$, and $\{a_i\} \not\sim \{a_i'\}$.

So suppose we have a Cauchy sequence $\{b_i\}$. We want to find an equivalent sequence $\{a_i\}$ satisfying (1) and (2). To do this we use a simple lemma.

Lemma. *If* $x \in \mathbb{Q}$ *and* $|x|_p \leq 1$, *then for any* i *there exists an integer* $\alpha \in \mathbb{Z}$ *such that* $|\alpha - x|_p \leq p^{-i}$. *The integer* α *can be chosen in the set* $\{0, 1, 2, 3, \ldots, p^i - 1\}$.

PROOF OF LEMMA. Let $x = a/b$ be written in lowest terms. Since $|x|_p \leq 1$, it follows that p does not divide b, and hence b and p^i are relatively prime. So we can find integers m and n such that: $mb + np^i = 1$. Let $\alpha = am$. The idea is that mb differs from 1 by a p-adically small amount, so that m is a good approximation to $1/b$, and so am is a good approximation to $x = a/b$. More precisely, we have:

$$|\alpha - x|_p = |am - (a/b)|_p = |a/b|_p \, |mb - 1|_p$$
$$\leq |mb - 1|_p = |np^i|_p = |n|_p/p^i \leq 1/p^i.$$

Finally, we can add a multiple of p^i to the integer α to get an integer between 0 and p^i for which $|\alpha - x|_p \leq p^{-i}$ still holds. The lemma is proved. □

Returning to the proof of the theorem, we look at our sequence $\{b_i\}$, and, for every $j = 1, 2, 3, \ldots$, let $N(j)$ be a natural number such that $|b_i - b_{i'}|_p \leq p^{-j}$ whenever $i, i' \geq N(j)$. (We may take the sequence $N(j)$ to be strictly increasing with j; in particular, $N(j) \geq j$.) Notice that $|b_i|_p \leq 1$ if $i \geq N(1)$, because for all $i' \geq N(1)$

$$|b_i|_p \leq \max(|b_{i'}|_p, |b_i - b_{i'}|_p)$$
$$\leq \max(|b_{i'}|_p, 1/p),$$

and $|b_{i'}|_p \to |a|_p \leq 1$ as $i' \to \infty$.

We now use the lemma to find a sequence of integers a_j, where $0 \leq a_j < p_j$, such that

$$|a_j - b_{N(j)}|_p \leq 1/p^j.$$

I claim that $\{a_j\}$ is the required sequence. It remains to show that $a_{j+1} \equiv a_j$ (mod p^j) and that $\{b_i\} \sim \{a_j\}$.

The first assertion follows because

$$|a_{j+1} - a_j|_p = |a_{j+1} - b_{N(j+1)} + b_{N(j+1)} - b_{N(j)} - (a_j - b_{N(j)})|_p$$
$$\leq \max(|a_{j+1} - b_{N(j+1)}|_p, |b_{N(j+1)} - b_{N(j)}|_p, |a_j - b_{N(j)}|_p)$$
$$\leq \max(1/p^{j+1}, 1/p^j, 1/p^j)$$
$$= 1/p^j.$$

The second assertion follows because, given any j, for $i \geq N(j)$ we have

$$|a_i - b_i|_p = |a_i - a_j + a_j - b_{N(j)} - (b_i - b_{N(j)})|_p$$
$$\leq \max(|a_i - a_j|_p, |a_j - b_{N(j)}|_p, |b_i - b_{N(j)}|_p)$$
$$\leq \max(1/p^j, 1/p^j, 1/p^j)$$
$$= 1/p^j.$$

Hence $|a_i - b_i|_p \to 0$ as $i \to \infty$. The theorem is proved. □

What if our p-adic number a does not satisfy $|a|_p \le 1$? Then we can multiply a by a power p^m of p (namely, by the power of p which equals $|a|_p$), to get a p-adic number $a' = ap^m$ which does satisfy $|a'|_p \le 1$. Then a' is represented by a sequence $\{a_i'\}$ as in the theorem, and $a = a'p^{-m}$ is represented by the sequence $\{a_i\}$ in which $a_i = a_i'p^{-m}$.

It is now convenient to write all the a_i' in the sequence for a' to the base p, i.e.,

$$a_i' = b_0 + b_1 p + b_2 p^2 + \cdots + b_{i-1} p^{i-1},$$

where the b's are all "digits," i.e., integers in $\{0, 1, \ldots, p-1\}$. Our condition $a_i' \equiv a_{i+1}' \pmod{p^i}$ precisely means that

$$a_{i+1}' = b_0 + b_1 p + b_2 p^2 + \cdots + b_{i-1} p^{i-1} + b_i p^i,$$

where the digits b_0 through b_{i-1} are all the *same* as for a_i'. Thus, a' can be thought of intuitively as a number, written to the base p, which extends infinitely far to the right, i.e., we add a new digit each time we pass from a_i' to a_{i+1}'.

Our original a can then be thought of as a base p decimal number which has only finitely many digits "to the right of the decimal point" (i.e., corresponding to negative powers of p, but actually written starting from the left) but has infinitely many digits for positive powers of p:

$$a = \frac{b_0}{p^m} + \frac{b_1}{p^{m-1}} + \cdots + \frac{b_{m-1}}{p} + b_m + b_{m+1} p + b_{m+2} p^2 + \cdots.$$

Here for the time being the expression on the right is only shorthand for the sequence $\{a_i\}$, where $a_i = b_0 p^{-m} + \cdots + b_{i-1} p^{i-1-m}$, that is, a convenient way of thinking of the sequence $\{a_i\}$ all at once. We'll soon see that this equality is in a precise sense "real" equality. This equality is called the "p-adic expansion" of a.

We let $\mathbb{Z}_p = \{a \in \mathbb{Q}_p \mid |a|_p \le 1\}$. This is the set of all numbers in $\mathbb{Q}_p$ whose p-adic expansion involves no negative powers of p. An element of $\mathbb{Z}_p$ is called a "p-adic integer." (From now on, to avoid confusion, when we mean an old-fashioned integer in $\mathbb{Z}$, we'll say "rational integer.") The sum, difference, and product of two elements of $\mathbb{Z}_p$ is in $\mathbb{Z}_p$, so $\mathbb{Z}_p$ is what's called a "subring" of the field $\mathbb{Q}_p$.

If $a, b \in \mathbb{Q}_p$, we write $a \equiv b \pmod{p^n}$ if $|a - b|_p \le p^{-n}$, or equivalently, $(a - b)/p^n \in \mathbb{Z}_p$, i.e., if the first nonzero digit in the p-adic expansion of $a - b$ occurs no sooner than the p^n-place. If a and b are not only in $\mathbb{Q}_p$ but are actually in $\mathbb{Z}$ (i.e., are rational integers), then this definition agrees with the earlier definition of $a \equiv b \pmod{p^n}$.

We define $\mathbb{Z}_p^\times$ as $\{x \in \mathbb{Z}_p \mid 1/x \in \mathbb{Z}_p\}$, or equivalently as $\{x \in \mathbb{Z}_p \mid x \not\equiv 0 \pmod{p}\}$, or equivalently as $\{x \in \mathbb{Z}_p \mid |x|_p = 1\}$. A p-adic integer in $\mathbb{Z}_p^\times$—i.e., whose first digit is nonzero—is sometimes called a "p-adic unit."

Now let $\{b_i\}_{i=-m}^{\infty}$ be any sequence of *p*-adic integers. Consider the sum

$$S_N = \frac{b_{-m}}{p^m} + \frac{b_{-m+1}}{p^{m-1}} + \cdots + b_0 + b_1 p + b_2 p^2 + \cdots + b_N p^N.$$

This sequence of partial sums is clearly Cauchy: if $M > N$, then $|S_N - S_M|_p < 1/p^N$. It therefore converges to an element in $\mathbb{Q}_p$. As in the case of infinite series of real numbers, we define $\sum_{i=-m}^{\infty} b_i p^i$ to be this limit in $\mathbb{Q}_p$.

More generally, if $\{c_i\}$ is *any* sequence of *p*-adic numbers such that $|c_i|_p \to 0$ as $i \to \infty$, the sequence of partial sums $S_N = c_1 + c_2 + \cdots + c_N$ converges to a limit, which we denote $\sum_{i=1}^{\infty} c_i$. This is because: $|S_M - S_N|_p = |c_{N+1} + c_{N+2} + \cdots + c_M|_p \leq \max(|c_{N+1}|_p, |c_{N+2}|_p, \cdots, |c_M|_p)$ which $\to 0$ as $N \to \infty$. Thus, *p*-adic infinite series are easier to check for convergence than infinite series of real numbers. *A series converges in $\mathbb{Q}_p$ if and only if its terms approach zero.* There is nothing like the harmonic series $1 + \frac{1}{2} + \frac{1}{3} + \frac{1}{4} + \cdots$ of real numbers, which diverges even though its terms approach 0. Recall that the reason for this is that $|\ |_p$ of a sum is bounded by the *maximum* (rather than just the sum) of the $|\ |_p$ of the summands when $p \neq \infty$, i.e., when $|\ |_p$ is non-Archimedean.

Returning now to *p*-adic expansions, we see that the infinite series on the right in the definition of the *p*-adic expansion

$$\frac{b_0}{p^m} + \frac{b_1}{p^{m-1}} + \cdots + \frac{b_{m-1}}{p} + b_m + b_{m+1} p + b_{m+2} p^2 + \cdots$$

(here $b_i \in \{0, 1, 2, \ldots, p-1\}$) converges to a, and so the equality can be taken in the sense of the sum of an infinite series.

Note that the uniqueness assertion in Theorem 2 is something we don't have in the Archimedean case. Namely, terminating decimals can also be represented by decimals with repeating 9s: $1 = 0.9999\cdots$. But if two *p*-adic expansions converge to the same number in $\mathbb{Q}_p$, then they are the same, i.e., all of their digits are the same.

One final remark. Instead of $\{0, 1, 2, \ldots, p-1\}$ we could have chosen any other set $S = \{\alpha_0, \alpha_1, \alpha_2, \ldots, \alpha_{p-1}\}$ of *p*-adic integers having the property that $\alpha_i \equiv i \pmod{p}$ for $i = 0, 1, 2, \ldots, p-1$, and could then have defined our *p*-adic expansion to be of the form $\sum_{i=-m}^{\infty} b_i p^i$, where now the "digits" b_i are in the set S rather than in the set $\{0, 1, \ldots, p-1\}$. For most purposes, the set $\{0, 1, \ldots, p-1\}$ is the most convenient. But there is another set S, the so-called "Teichmüller representatives" (see Exercise 13 below), which is in some ways an even more natural choice.

5. Arithmetic in $\mathbb{Q}_p$

The mechanics of adding, subtracting, multiplying, and dividing *p*-adic numbers is very much like the corresponding operations on decimals which we learn to do in about the third grade. The only difference is that the "carrying," "borrowing," "long multiplication," etc. go from left to right rather than right to left. Here are a few examples in $\mathbb{Q}_7$:

$$
\begin{array}{r}
3 + 6 \times 7 + 2 \times 7^2 + \cdots \\
\times\; 4 + 5 \times 7 + 1 \times 7^2 + \cdots \\
\hline
5 + 4 \times 7 + 4 \times 7^2 + \cdots \\
1 \times 7 + 4 \times 7^2 + \cdots \\
3 \times 7^2 + \cdots \\
\hline
5 + 5 \times 7 + 4 \times 7^2 + \cdots
\end{array}
$$

$$
\begin{array}{r}
2 \times 7^{-1} + 0 \times 7^0 + 3 \times 7^1 + \cdots \\
-\; 4 \times 7^{-1} + 6 \times 7^0 + 5 \times 7^1 + \cdots \\
\hline
5 \times 7^{-1} + 0 \times 7^0 + 4 \times 7^1 + \cdots
\end{array}
$$

$$
\begin{array}{r}
5 + 1 \times 7 + 6 \times 7^2 + \cdots \\
3 + 5 \times 7 + 1 \times 7^2 + \cdots \overline{\big)\, 1 + 2 \times 7 + 4 \times 7^2 + \cdots} \\
1 + 6 \times 7 + 1 \times 7^2 + \cdots \\
\hline
3 \times 7 + 2 \times 7^2 + \cdots \\
3 \times 7 + 5 \times 7^2 + \cdots \\
\hline
4 \times 7^2 + \cdots \\
4 \times 7^2 + \cdots \\
\hline
\end{array}
$$

As another example, let's try to extract $\sqrt{6}$ in $\mathbb{Q}_5$, i.e., we want to find $a_0, a_1, a_2, \ldots, 0 \le a_i \le 4$, such that

$$(a_0 + a_1 \times 5 + a_2 \times 5^2 + \cdots)^2 = 1 + 1 \times 5.$$

Comparing coefficients of $1 = 5^0$ on both sides gives $a_0{}^2 \equiv 1 \pmod 5$, and hence $a_0 = 1$ or 4. Let's take $a_0 = 1$. Then comparing coefficients of 5 on both sides gives $2a_1 \times 5 \equiv 1 \times 5 \pmod{5^2}$, so that $2a_1 \equiv 1 \pmod 5$, and hence $a_1 = 3$. At the next step we have:

$$1 + 1 \times 5 \equiv (1 + 3 \quad 5 + a_2 \times 5^2)^2 \equiv 1 + 1 \times 5 + 2a_2 \times 5^2 \pmod{5^3}.$$

Hence $2a_2 \equiv 0 \pmod 5$, and $a_2 = 0$. Proceeding in this way, we get a series

$$a = 1 + 3 \times 5 + 0 \times 5^2 + 4 \times 5^3 + a_4 \times 5^4 + a_5 \times 5^5 + \cdots$$

where each a_i after a_0 is uniquely determined.

But remember that we had two choices for a_0, namely 1 and 4. What if we had chosen 4 instead of 1? We would have gotten

$$
\begin{aligned}
-a = {} & 4 + 1 \times 5 + 4 \times 5^2 + 0 \times 5^3 \\
& + (4 - a_4) \times 5^4 + (4 - a_5) \times 5^5 + \cdots.
\end{aligned}
$$

The fact that we had two choices for a_0, and then, once we chose a_0, only a single possibility for $a_1, a_2, a_3, \ldots$, merely reflects the fact that a nonzero element in a field like $\mathbb{Q}$ or $\mathbb{R}$ or $\mathbb{Q}_p$ always has exactly two square roots in the field if it has any.

Do all numbers in $\mathbb{Q}_5$ have square roots? We saw that 6 does, what about 7? If we had

$$(a_0 + a_1 \times 5 + \cdots)^2 = 2 + 1 \times 5,$$

it would follow that $a_0{}^2 \equiv 2 \pmod 5$. But this is impossible, as we see by checking the possible values $a_0 = 0, 1, 2, 3, 4$. For a more systematic look at square roots in $\mathbb{Q}_p$, see Exercises 6–12.

This method of solving the equation $x^2 - 6 = 0$ in $\mathbb{Q}_5$—by solving the congruence $a_0{}^2 - 6 \equiv 0 \pmod 5$ and then solving for the remaining a_i in a step-by-step fashion—is actually quite general, as shown by the following important "lemma." This form of the lemma was apparently first given in Serge Lang's Ph.D. thesis in 1952 (*Annals of Mathematics*, Vol. 55, p. 380).

Theorem 3 (Hensel's lemma). *Let $F(x) = c_0 + c_1 x + \cdots + c_n x^n$ be a polynomial whose coefficients are p-adic integers. Let $F'(x) = c_1 + 2c_2 x + 3c_3 x^2 + \cdots + nc_n x^{n-1}$ be the derivative of $F(x)$. Let a_0 be a p-adic integer such that $F(a_0) \equiv 0 \pmod p$ and $F'(a_0) \not\equiv 0 \pmod p$. Then there exists a unique p-adic integer a such that*

$$F(a) = 0 \quad \text{and} \quad a \equiv a_0 \pmod p.$$

(*Note*: In the special case treated above, we had $F(x) = x^2 - 6$, $F'(x) = 2x$, $a_0 = 1$.)

PROOF OF HENSEL'S LEMMA. I claim that there exists a unique sequence of rational integers $a_1, a_2, a_3, \ldots$ such that for all $n \geq 1$:

(1) $F(a_n) \equiv 0 \pmod{p^{n+1}}$.
(2) $a_n \equiv a_{n-1} \pmod{p^n}$.
(3) $0 \leq a_n < p^{n+1}$.

We prove that such a_n exist and are unique by induction on n.

If $n = 1$, first let $\tilde{a}_0$ be the unique integer in $\{0, 1, \ldots, p - 1\}$ which is congruent to $a_0 \bmod p$. Any a_1 satisfying (2) and (3) must be of the form $\tilde{a}_0 + b_1 p$, where $0 \leq b_1 \leq p - 1$. Now, looking at $F(\tilde{a}_0 + b_1 p)$, we expand the polynomial, remembering that we only need congruence to $0 \bmod p^2$, so that any terms divisible by p^2 may be ignored:

$$F(a_1) = F(\tilde{a}_0 + b_1 p) = \sum c_i(\tilde{a}_0 + b_1 p)^i$$
$$= \sum (c_i \tilde{a}_0{}^i + ic_i \tilde{a}_0{}^{i-1} b_1 p + \text{terms divisible by } p^2)$$
$$\equiv \sum c_i \tilde{a}_0{}^i + \left(\sum ic_i \tilde{a}_0{}^{i-1}\right) b_1 p \pmod{p^2}$$
$$= F(\tilde{a}_0) + F'(\tilde{a}_0) b_1 p.$$

(Note the similarity to the first order Taylor series approximation in calculus: $F(x + h) = F(x) + F'(x)h + \text{higher order terms.})$ Since $F(a_0) \equiv 0 \pmod p$ by assumption, we can write $F(\tilde{a}_0) \equiv \alpha p \pmod{p^2}$ for some $\alpha \in \{0, 1, \ldots, p - 1\}$. So in order to get $F(a_1) \equiv 0 \pmod{p^2}$ we must get $\alpha p + F'(\tilde{a}_0) b_1 p \equiv 0 \pmod{p^2}$, i.e., $\alpha + F'(\tilde{a}_0) b_1 \equiv 0 \pmod p$. But, since $F'(a_0) \not\equiv 0 \pmod p$ by assumption, this equation can always be solved for the unknown b_1. Namely, using the lemma in the proof of Theorem 2, we choose $b_1 \in \{0, 1, \ldots, p - 1\}$ so that $b_1 \equiv -\alpha/F'(\tilde{a}_0) \pmod p$. Clearly this $b_1 \in \{0, 1, \ldots, p - 1\}$ is uniquely determined by this condition.

Now, to proceed with the induction, suppose we already have $a_1, a_2, \ldots,$ a_{n-1}. We want to find a_n. By (2) and (3), we need $a_n = a_{n-1} + b_n p^n$ with $b_n \in \{0, 1, \ldots, p - 1\}$. We expand $F(a_{n-1} + b_n p^n)$ as we did before in the case $n = 1$, only this time we ignore terms divisible by p^{n+1}. This gives us:

$$F(a_n) = F(a_{n-1} + b_n p^n) \equiv F(a_{n-1}) + F'(a_{n-1}) b_n p^n \pmod{p^{n+1}}.$$

Since $F(a_{n-1}) \equiv 0 \pmod{p^n}$ by the induction assumption, we can write $F(a_{n-1}) \equiv \alpha' p^n \pmod{p^{n+1}}$, and our desired condition $F(a_n) \equiv 0 \pmod{p^{n+1}}$ now becomes

$$\alpha' p^n + F'(a_{n-1}) b_n p^n \equiv 0 \pmod{p^{n+1}}, \quad \text{i.e.,} \quad \alpha' + F'(a_{n-1}) b_n \equiv 0 \pmod{p}.$$

Now, since $a_{n-1} \equiv a_0 \pmod{p}$, it easily follows that $F'(a_{n-1}) \equiv F'(a_0) \not\equiv 0$ $\pmod{p}$, and we can find the required $b_n \in \{0, 1, \ldots, p - 1\}$ proceeding exactly as in the case of b_1, i.e., solving $b_n \equiv -\alpha'/F'(a_{n-1}) \pmod{p}$. This completes the induction step, and hence the proof of the claim.

The theorem follows immediately from the claim. We merely let $a = \tilde{a}_0 + b_1 p + b_2 p^2 + \cdots$. Since for all n we have $F(a) \equiv F(a_n) \equiv 0 \pmod{p^{n+1}}$, it follows that the p-adic number $F(a)$ must be 0. Conversely, any $a = \tilde{a}_0 + b_1 p + b_2 p^2 + \cdots$ gives a sequence of a_n as in the claim, and the uniqueness of that sequence implies the uniqueness of the a. Hensel's lemma is proved. $\square$

Hensel's lemma is often called the p-adic Newton's lemma because the approximation technique used to prove it is essentially the same as Newton's

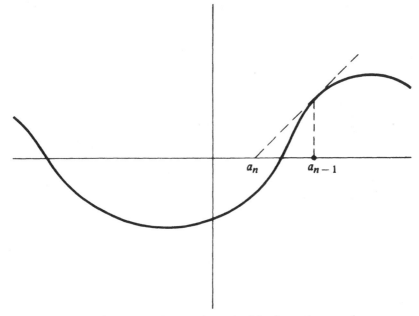

Figure I.1.   Newton's method in the real case

method for finding a *real* root of a polynomial equation with real coefficients. In Newton's method in the real case, (see Figure I.1), if $f'(a_{n-1}) \neq 0$, we take

$$a_n = a_{n-1} - \frac{f(a_{n-1})}{f'(a_{n-1})}.$$

The correction term $-f(a_{n-1})/f'(a_{n-1})$ is a lot like the formula for the "correction term" in the proof of Hensel's lemma:

$$b_n p^n \equiv -\frac{\alpha' p^n}{F'(a_{n-1})} \equiv -\frac{F(a_{n-1})}{F'(a_{n-1})} \pmod{p^{n+1}}.$$

In one respect the *p*-adic Newton's method (Hensel's lemma) is much better than Newton's method in the real case. In the *p*-adic case, it's *guaranteed* to converge to a root of the polynomial. In the real case, Newton's method *usually* converges, but not always. For example, if you take $f(x) = x^3 - x$ and make the unfortunate choice $a_0 = 1/\sqrt{5}$, you get:

$$a_1 = 1/\sqrt{5} - [1/5\sqrt{5} - 1/\sqrt{5}]/(3/5 - 1)$$
$$= 1/\sqrt{5}[1 - (1/5 - 1)(3/5 - 1)] = -1/\sqrt{5};$$
$$a_2 = 1/\sqrt{5}; \quad a_3 = -1/\sqrt{5}, \quad \text{etc.}$$

(See Figure I.2.) Such perverse silliness is impossible in $\mathbb{Q}_p$.

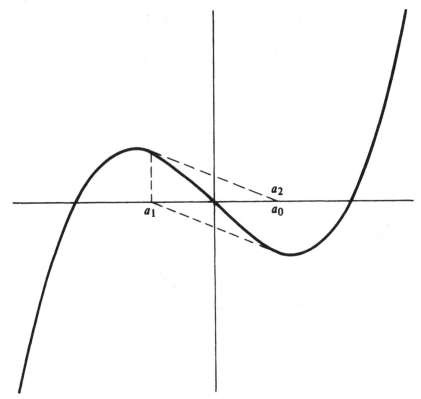

Figure I.2. Failure of Newton's method in the real case

EXERCISES

1. If $a \in \mathbb{Q}_p$ has p-adic expansion $a_{-m}p^{-m} + a_{-m+1}p^{-m+1} + \cdots + a_0 + a_1 p + \cdots$, what is the p-adic expansion of $-a$?

2. Find the p-adic expansion of:

   (i) $(6 + 4 \times 7 + 2 \times 7^2 + 1 \times 7^3 + \cdots)(3 + 0 \times 7 + 0 \times 7^2 + 6 \times 7^3 + \cdots)$ in $\mathbb{Q}_7$ to 4 digits

   (ii) $1/(3 + 2 \times 5 + 3 \times 5^2 + 1 \times 5^3 + \cdots)$ in $\mathbb{Q}_5$ to 4 digits

   (iii) $9 \times 11^2 - (3 \times 11^{-1} + 2 + 1 \times 11^1 + 3 \times 11^2 + \cdots)$ in $\mathbb{Q}_{11}$ to 4 digits

   (iv) $2/3$ in $\mathbb{Q}_2$ (v) $-1/6$ in $\mathbb{Q}_7$ (vi) $1/10$ in $\mathbb{Q}_{11}$

   (vii) $-9/16$ in $\mathbb{Q}_{13}$ (viii) $1/1000$ in $\mathbb{Q}_5$ (ix) $6!$ in $\mathbb{Q}_3$

   (x) $1/3!$ in $\mathbb{Q}_3$ (xi) $1/4!$ in $\mathbb{Q}_2$ (xii) $1/5!$ in $\mathbb{Q}_5$

3. Prove that the p-adic expansion of a nonzero $a \in \mathbb{Q}_p$ terminates (i.e., $a_i = 0$ for all i greater than some N) if and only if a is a *positive* rational number whose denominator is a power of p.

4. Prove that the p-adic expansion of $a \in \mathbb{Q}_p$ has repeating digits from some point on (i.e., $a_{i+r} = a_i$ for some r and for all i greater than some N) if and only if $a \in \mathbb{Q}$.

5. What is the cardinality of $\mathbb{Z}_p$? Prove your answer.

6. Prove the following generalization of Hensel's lemma: Let $F(x)$ be a polynomial with coefficients in $\mathbb{Z}_p$. If $a_0 \in \mathbb{Z}_p$ satisfies $F'(a_0) \equiv 0 \pmod{p^M}$ but $F'(a_0) \not\equiv 0 \pmod{p^{M+1}}$, and if $F(a_0) \equiv 0 \pmod{p^{2M+1}}$, then there is a unique $a \in \mathbb{Z}_p$ such that $F(a) = 0$ and $a \equiv a_0 \pmod{p^{M+1}}$.

7. Use your proof in Exercise 6 to find a square root of -7 in $\mathbb{Q}_2$ to 5 digits.

8. Which of the following 11-adic numbers have square roots in $\mathbb{Q}_{11}$?

   (i) 5   (ii) 7   (iii) -7

   (iv) $5 + 3 \times 11 + 9 \times 11^2 + 1 \times 11^3$

   (v) $3 \times 11^{-2} + 6 \times 11^{-1} + 3 + 0 \times 11 + 7 \times 11^2$

   (vi) $3 \times 11^{-1} + 6 + 3 \times 11 + 0 \times 11^2 + 7 \times 11^3$

   (vii) 1×11^7 (viii) $7 - 6 \times 11^2$

   (ix) $5 \times 11^{-2} + \sum_{n=0}^{\infty} n \times 11^n$.

9. Compute $\pm\sqrt{-1}$ in $\mathbb{Q}_5$ and $\pm\sqrt{-3}$ in $\mathbb{Q}_7$ to 4 digits.

10. For which $p = 2, 3, 5, 7, 11, 13, 17, 19$ does -1 have a square root in $\mathbb{Q}_p$?

11. Let p be any prime besides 2. Suppose $\alpha \in \mathbb{Q}_p$ and $|\alpha|_p = 1$. Describe a test for whether α has a square root in $\mathbb{Q}_p$. What about if $|\alpha|_p \neq 1$? Prove that there exist four numbers $\alpha_1, \alpha_2, \alpha_3, \alpha_4, \in \mathbb{Q}_p$ such that for all nonzero $\alpha \in \mathbb{Q}_p$ exactly one of the numbers $\alpha_1\alpha$, $\alpha_2\alpha$, $\alpha_3\alpha$, $\alpha_4\alpha$ has a square root. (In the case when p is replaced by ∞ and $\mathbb{Q}_p$ by $\mathbb{R}$, there are *two* numbers, for example ± 1 will do, such that for every nonzero $\alpha \in \mathbb{R}$ exactly one of the numbers $1 \cdot \alpha$ and $-1 \cdot \alpha$ has a square root in $\mathbb{R}$.)

12. The same as Exercise 11 when $p = 2$, except that now there will be *eight* numbers $\alpha_1, \ldots, \alpha_8 \in \mathbb{Q}_2$ such that for all nonzero $\alpha \in \mathbb{Q}_2$ exactly one of the

19

numbers $\alpha_1\alpha, \ldots, \alpha_8\alpha$ has a square root in $\mathbb{Q}_2$. Find such $\alpha_1, \ldots, \alpha_8$ (the choice of them is *not* unique, of course).

13. Find all 4 fourth roots of 1 in $\mathbb{Q}_5$ to four digits. Prove that $\mathbb{Q}_p$ always contains p solutions $a_0, a_1, \ldots, a_{p-1}$ to the equation $x^p - x = 0$, where $a_i \equiv i \pmod{p}$. These p numbers are called the "Teichmüller representatives" of $\{0, 1, 2, \ldots, p - 1\}$ and are sometimes used as a set of p-adic digits instead of $\{0, 1, 2, \ldots, p - 1\}$. If $p > 2$, which Teichmüller representatives are rational?

14. Prove the following "Eisenstein irreducibility criterion" for a polynomial $f(x) = a_0 + a_1x + \cdots + a_nx^n$ with coefficients $a_i \in \mathbb{Z}_p$: If $a_i \equiv 0 \pmod{p}$ for $i = 0, 1, 2, \ldots, n - 1$, if $a_n \not\equiv 0 \pmod{p}$, and if $a_0 \not\equiv 0 \pmod{p^2}$, then $f(x)$ is irreducible over $\mathbb{Q}_p$, i.e., it cannot be written as a product of two lower degree polynomials with coefficients in $\mathbb{Q}_p$.

15. If $p > 2$, use Exercise 14 to show that 1 has no pth root other than 1 in $\mathbb{Q}_p$. Prove that if $p > 2$, then the only roots of 1 in $\mathbb{Q}_p$ are the nonzero Teichmüller representatives; and in $\mathbb{Q}_2$ the only roots of 1 are ± 1.

16. Prove that the infinite sum $1 + p + p^2 + p^3 + \cdots$ converges to $1/(1 - p)$ in $\mathbb{Q}_p$. What about $1 - p + p^2 - p^3 + p^4 - p^5 + \cdots$? What about $1 + (p - 1)p + p^2 + (p - 1)p^3 + p^4 + (p - 1)p^5 + \cdots$?

17. Show that (a) every element $x \in \mathbb{Z}_p$ has a unique expansion of the form $x = a_0 + a_1(-p) + a_2(-p)^2 + \cdots + a_n(-p)^n + \cdots$, with $a_i \in \{0, 1, \ldots, p - 1\}$, and (b) this expansion terminates if and only if $x \in \mathbb{Z}$.

18. Suppose that n is a (positive or negative) integer not divisible by p, and let $\alpha \equiv 1 \pmod{p}$. Show that α has an nth root in $\mathbb{Q}_p$. Give a counter-example if $n = p$. Show that α has a pth root if $\alpha \equiv 1 \pmod{p^2}$ and $p \neq 2$.

19. Let $\alpha \in \mathbb{Z}_p$. Prove that $\alpha^{p^M} \equiv \alpha^{p^{M-1}} \pmod{p^M}$ for $M = 1, 2, 3, 4, \ldots$. Prove that the sequence $\{\alpha^{p^M}\}$ approaches a limit in $\mathbb{Q}_p$, and that this limit is the Teichmüller representative congruent to α mod p.

20. Prove that $\mathbb{Z}_p$ is sequentially compact, i.e., every sequence of p-adic integers has a convergent subsequence.

21. Define matrices with entries in $\mathbb{Q}_p$, their sums, products, and determinants exactly as in the case of the reals. Let $M = \{r \times r$ matrices with entries in $\mathbb{Z}_p\}$, let $M^\times = \{A \in M \mid A$ has an inverse in $M\}$ (it's not hard to see that this is equivalent to: $\det A \in \mathbb{Z}_p^\times$), and let $pM = \{A \in M \mid A = pB$ with $B \in M\}$. If $A \in M^\times$ and $B \in pM$, prove that there exists a unique $X \in M^\times$ such that: $X^2 - AX + B = 0$.

CHAPTER II

p-adic interpolation of the Riemann zeta-function

This chapter is logically independent of the following chapters, and is presented at this point in the middle of our ascent to Ω as a plateau in the level of abstraction—namely, everything in this chapter still takes place in the fields $\mathbb{Q}$, $\mathbb{Q}_p$, and $\mathbb{R}$.

The Riemann ζ-function is defined as a function of real numbers greater than 1 by

$$\zeta(s) \underset{\text{def}}{=} \sum_{n=1}^{\infty} \frac{1}{n^s}.$$

It is easy to see (by comparison with the integral $\int_1^\infty (dx/x^s) = 1/(s-1)$ for fixed $s > 1$) that this sum converges when $s > 1$.

Let p be any prime number. The purpose of this chapter is to show that the numbers $\zeta(2k)$ for $k = 1, 2, 3, \ldots$ have a "p-adic continuity property." More precisely, consider the set of numbers

$$f(2k) = (1 - p^{2k-1}) \frac{c_k}{\pi^{2k}} \zeta(2k), \quad \text{where } c_k = (-1)^k \frac{(2k-1)!}{2^{2k-1}},$$

as $2k$ runs through all positive even integers in the same congruence class $\mod(p-1)$. It turns out that $f(2k)$ is always a rational number. Moreover, if two such values of $2k$ are close p-adically (i.e., their difference is divisible by a high power of p), then we shall see that the corresponding $f(2k)$ are also p-adically close. (We must also assume that $2k$ is not divisible by $p - 1$.) This means that the function f can be extended in a unique way from integers to p-adic integers so that the resulting function is a *continuous function of a p-adic variable with values in* $\mathbb{Q}_p$. ("Continuous function" means, as in the real case, that whenever a sequence of p-adic integers $\{x_n\}$ approaches x p-adically, $\{f(x_n)\}$ approaches $f(x)$ p-adically.)

This is what is meant by *p*-adic "interpolation." The process is analogous to the classical procedure for, say, defining the function $f(x) = a^x$ (where a is a fixed positive real number): first define $f(x)$ for fractional x; then prove that nearby fractional values of x give nearby values of a^x; and, finally, define a^x for x irrational to be the limit of a^{x_n} for any sequence of rational numbers x_n which approach x.

Notice that a function f on the set S of, for example, positive even integers can be extended in at most one way to a continuous function on $\mathbb{Z}_p$ (assume $p \neq 2$). This is because S is "dense" in $\mathbb{Z}_p$—any $x \in \mathbb{Z}_p$ can be written as a limit of positive even integers x_n. If f is to be continuous, we must have $f(x) = \lim_{n \to \infty} f(x_n)$. In the real case, while the rational numbers *are* dense in $\mathbb{R}$, the set S is not. It makes no sense to talk of "the" continuous real-valued function which interpolates a function on the positive even integers; there are always infinitely many such functions. (However, there might be a unique real-valued continuous interpolating function which has additional convenient properties: for example, the gamma-function $\Gamma(x + 1)$ interpolates $k!$ when $x = k$ is a nonnegative integer, it satisfies $\Gamma(x + 1) = x\Gamma(x)$ for all real x, and its logarithm is a convex function for $x > 0$; the gamma-function is uniquely characterized by these properties.)

1. A formula for $\zeta(2k)$

The kth Bernoulli number B_k is defined as $k!$ times the kth coefficient in the Taylor series for

$$\frac{t}{e^t - 1} = \frac{1}{1 + t/2! + t^2/3! + t^3/4! + \cdots + t^n/(n + 1)! + \cdots}$$

$$\underset{\text{def}}{=} \sum_{k=0}^{\infty} B_k t^k / k!.$$

The first few B_k are:

$$B_0 = 1, \quad B_1 = -1/2, \quad B_2 = 1/6, \quad B_3 = 0,$$
$$B_4 = -1/30, \quad B_5 = 0, \quad B_6 = 1/42, \ldots.$$

We now derive the formula:

$$\zeta(2k) = (-1)^k \pi^{2k} \frac{2^{2k-1}}{(2k - 1)!} \left(-\frac{B_{2k}}{2k} \right) \quad \text{for } k = 1, 2, 3, \ldots.$$

Recall the definition of the "hyperbolic sine," abbreviated sinh (and pronounced "sinch"):

$$\sinh x = \frac{e^x - e^{-x}}{2}.$$

It is equal to its Taylor series

$$\sinh x = x + \frac{x^3}{3!} + \frac{x^5}{5!} + \cdots + \frac{x^{2k+1}}{(2k + 1)!} + \cdots,$$

obtained by averaging the series for e^x and $-e^{-x}$. Note that this Taylor series is the same as that for sin x, except without the alternation of sign.

Proposition. *For all real numbers x, the infinite product*

$$\pi x \prod_{n=1}^{\infty} \left(1 + \frac{x^2}{n^2}\right)$$

converges and equals $\sinh(\pi x)$.

PROOF. Convergence of the infinite product is immediate from the logarithm test:

$$\sum_{n=1}^{\infty} \left| \log\left(1 + \frac{x^2}{n^2}\right) \right| \le \sum_{n=1}^{\infty} \frac{x^2}{n^2} < \infty \quad \text{for all } x.$$

We start by deriving the infinite product for sin x.

Lemma. *Let $n = 2k + 1$ be a positive odd integer. Then we can write*

$$\sin(nx) = P_n (\sin x)$$
$$\cos(nx) = \cos x \, Q_{n-1} (\sin x)$$

where P_n (respectively Q_{n-1}) is a polynomial of degree at most n (respectively $n - 1$) with integer coefficients.

PROOF OF LEMMA. We use induction on k. The lemma is trivial for $k = 0$ (i.e., $n = 1$). Suppose it holds for $k - 1$. Then

$$
\begin{aligned}
\sin[(2k + 1)x] &= \sin[(2k - 1)x + 2x] \\
&= \sin(2k - 1)x \cos 2x + \cos(2k - 1)x \sin 2x \\
&= P_{2k-1} (\sin x)(1 - 2 \sin^2 x) \\
&\quad + \cos x Q_{2k-2} (\sin x)2 \sin x \cos x,
\end{aligned}
$$

which is of the required form $P_{2k+1} (\sin x)$. The proof that $\cos(2k + 1)x = \cos x \, Q_{2k} (\sin x)$ is completely similar, and will be left to the reader. $\square$

We now return to the proof of the proposition. Notice that, if we set $x = 0$ in sin $nx = P_n (\sin x)$, we find that P_n has constant term zero. Next, we take the derivative with respect to x of both sides of sin $nx = P_n (\sin x)$:

$$n \cos nx = P_n'(\sin x) \cos x.$$

Setting $x = 0$ here gives: $n = P_n'(0)$, i.e., the first coefficient of P_n is n. Thus, we may write:

$$\frac{\sin nx}{n \sin x} = \tilde{P}_{2k}(\sin x) = 1 + a_1 \sin x + a_2 \sin^2 x + \cdots$$

$$+ a_{2k} \sin^{2k} x \qquad (n = 2k + 1),$$

where the a_i are rational numbers. Note that for $x = \pm (\pi/n), \ldots, \pm (k\pi/n)$, the left-hand side vanishes. But the $2k$ values $y = \pm \sin(\pi/n), \pm \sin(2\pi/n), \ldots,$

$\pm \sin(k\pi/n)$ are distinct numbers at which the polynomial $\tilde{P}_{2k}(y)$ vanishes. Since $\tilde{P}_{2k}$ has degree $2k$ and constant term 1, we must have:

$$\tilde{P}_{2k}(y) = \left(1 - \frac{y}{\sin \pi/n}\right)\left(1 - \frac{y}{-\sin \pi/n}\right)\left(1 - \frac{y}{\sin 2\pi/n}\right)\left(1 - \frac{y}{-\sin 2\pi/n}\right)$$

$$\cdots \left(1 - \frac{y}{\sin k\pi/n}\right)\left(1 - \frac{y}{-\sin k\pi/n}\right)$$

$$= \prod_{r=1}^{k}\left(1 - \frac{y^2}{\sin^2 r\pi/n}\right).$$

Thus

$$\frac{\sin nx}{n \sin x} = \tilde{P}_{2k}(\sin x) = \prod_{r=1}^{k}\left(1 - \frac{\sin^2 x}{\sin^2 r\pi/n}\right).$$

Replacing x by $\pi x/n$ gives:

$$\frac{\sin \pi x}{n \sin(\pi x/n)} = \prod_{r=1}^{k}\left(1 - \frac{\sin^2(\pi x/n)}{\sin^2(\pi r/n)}\right).$$

Now take the limit of both sides as $n = 2k + 1 \to \infty$. The left-hand side approaches $(\sin \pi x)/\pi x$. For r small relative to n the rth term in the product approaches $1 - ((\pi x/n)/(\pi r/n))^2 = 1 - (x^2/r^2)$. It then follows that the product converges to $\prod_{r=1}^{\infty}(1 - (x^2/r^2))$. (The rigorous justification is straightforward, and will be left as an exercise below.)

We conclude:

$$\prod_{n=1}^{\infty}\left(1 - \frac{x^2}{n^2}\right) = \frac{\sin(\pi x)}{\pi x} = 1 - \frac{\pi^2 x^2}{3!} + \frac{\pi^4 x^4}{5!} - \frac{\pi^6 x^6}{7!} + \frac{\pi^8 x^8}{9!} - \cdots,$$

using the Taylor series for sine. But

$$\frac{\sinh(\pi x)}{\pi x} = 1 + \frac{\pi^2 x^2}{3!} + \frac{\pi^4 x^4}{5!} + \frac{\pi^6 x^6}{7!} + \frac{\pi^8 x^8}{9!} + \cdots.$$

If we multiply out the infinite product for $\sin(\pi x)/(\pi x)$, we get a minus sign precisely in those terms having an odd number of x^2/n^2 terms, i.e., precisely for the terms in the Taylor series for $\sin(\pi x)/(\pi x)$ having a minus sign. Thus, changing the sign in the infinite product has the effect of changing all of the $-$'s to $+$'s on the right, and we have the desired product expansion of the proposition. (For a "better" way of thinking of this last step, see Exercise 3 below.) ☐

We are now ready to prove:

Theorem 4.

$$\zeta(2k) = (-1)^k \pi^{2k} \frac{2^{2k-1}}{(2k-1)!}\left(-\frac{B_{2k}}{2k}\right).$$

PROOF. First take the logarithm of both sides of

$$\sinh(\pi x) = \pi x \prod_{n=1}^{\infty} \left(1 + \frac{x^2}{n^2}\right)$$

(for $x > 0$). On the left we get

$$\log \sinh(\pi x) = \log[(e^{\pi x} - e^{-\pi x})/2] = \log[(e^{\pi x}/2)(1 - e^{-2\pi x})]$$
$$= \log(1 - e^{-2\pi x}) + \pi x - \log 2.$$

On the right we get (for $0 < x < 1$)

$$\log \pi + \log x + \sum_{n=1}^{\infty} \log(1 + x^2/n^2) = \log \pi + \log x + \sum_{n=1}^{\infty} \sum_{k=1}^{\infty} (-1)^{k+1} \frac{x^{2k}}{kn^{2k}},$$

by the Taylor series for $\log(1 + x)$. Since this double series is absolutely convergent for $0 < x < 1$, we can interchange the order of summation and obtain the equality:

$$\log(1 - e^{-2\pi x}) + \pi x - \log 2 = \log \pi + \log x + \sum_{k=1}^{\infty} \left[(-1)^{k+1} \frac{x^{2k}}{k} \sum_{n=1}^{\infty} \frac{1}{n^{2k}}\right]$$

$$= \log \pi + \log x + \sum_{k=1}^{\infty} (-1)^{k+1} \frac{x^{2k}}{k} \zeta(2k).$$

We now take the derivative of both sides with respect to x. On the right we may differentiate term-by-term, since the resulting series is uniformly convergent in $0 < x < 1 - \varepsilon$ for any $\varepsilon > 0$. Thus,

$$\frac{2\pi e^{-2\pi x}}{1 - e^{-2\pi x}} + \pi = \frac{1}{x} + 2 \sum_{k=1}^{\infty} (-1)^{k+1} x^{2k-1} \zeta(2k).$$

Multiplying through by x and then substituting $x/2$ for x gives:

$$\frac{\pi x}{e^{\pi x} - 1} + \frac{\pi x}{2} = 1 + \sum_{k=1}^{\infty} \frac{(-1)^{k+1} \zeta(2k)}{2^{2k-1}} x^{2k}.$$

The left-hand side gives: $(\pi x)/2 + \sum_{k=0}^{\infty} B_k (\pi x)^k/k!$. Comparing coefficients of even powers of x gives: $\pi^{2k} B_{2k}/(2k)! = ((-1)^{k+1}/2^{2k-1}) \zeta(2k)$, which gives us the theorem. $\qquad \square$

As examples, we have

$$\zeta(2) = \frac{\pi^2}{6}, \qquad \zeta(4) = \frac{\pi^4}{90}, \qquad \zeta(6) = \frac{\pi^6}{945}.$$

The arrangement of the formula for $\zeta(2k)$ in the statement of Theorem 4 was deliberate. We think of the $(-B_{2k}/2k)$ as the "interesting" part, and the $(-1)^k \pi^{2k} 2^{2k-1}/(2k - 1)!$ as a nuisance factor. It is the interesting part that we end up interpolating p-adically. Some justification for taking $(-B_{2k}/2k)$ rather than the whole mess will be given later (§7). For now, let's remark that at least the π^{2k} factor has to be discarded when we interpolate the values *p*-adically, since transcendental real numbers cannot be considered *p*-adically in any reasonable way. (What could "*p*-adic ordinal" mean for them?)

2. *p*-adic interpolation of the function $f(s) = a^s$

This section will eventually play a role in the subsequent logical development. It is included at this time as a "dry run" in order to motivate certain features of the later *p*-adic interpolation which may otherwise seem somewhat idiosyncratic.

As mentioned before, if *a* is a fixed positive real number, the function $f(s) = a^s$ is defined as a continuous function of a real variable by first defining it on the set of rational numbers *s*, and then "interpolating" or "extending by continuity" to real numbers, each of which can be written as the limit of a sequence of rational numbers.

Now suppose that $a = n$ is a fixed positive integer. Consider *n* as an element in $\mathbb{Q}_p$. For every nonnegative integer *s*, the integer n^s belongs to $\mathbb{Z}_p$. Now the nonnegative integers are dense in $\mathbb{Z}_p$ in the same way as $\mathbb{Q}$ is dense in $\mathbb{R}$. In other words, every *p*-adic integer is the limit of a sequence of nonnegative integers (for example, the partial sums in its *p*-adic expansion). So we might try to extend $f(s) = n^s$ by continuity from nonnegative integers *s* to all *p*-adic integers *s*.

To do this, we must ask if n^s and $n^{s'}$ are close whenever the two nonnegative integers *s* and *s'* are close, for example, when $s' = s + p^N$ for some large *N*. A couple of examples show that this is not always the case:

(1) $n = p, s = 0$: $|n^s - n^{s'}|_p = |1 - p^{p^N}|_p = 1$ no matter what *N* is.
(2) $1 < n < p$: by Fermat's Little Theorem (see §III.1, especially the first paragraph of the proof of Theorem 9), we have $n \equiv n^p \pmod{p}$, and so $n \equiv n^p \equiv n^{p^2} \equiv n^{p^3} \equiv \cdots \equiv n^{p^N} \pmod{p}$; hence $n^s - n^{s + p^N} = n^s(1 - n^{p^N}) \equiv n^s(1 - n) \pmod{p}$; thus $|n^s - n^{s'}|_p = 1$ no matter what *N* is.

But the situation is not as bad as these examples make it seem. Let's choose *n* so that $n \equiv 1 \pmod{p}$, say $n = 1 + mp$. Let $|s' - s|_p \le 1/p^N$, so that $s' = s + s''p^N$ for some $s'' \in \mathbb{Z}$. Then we have (say $s' > s$)

$$|n^s - n^{s'}|_p = |n^s|_p|1 - n^{s' - s}|_p = |1 - n^{s' - s}|_p = |1 - (1 + mp)^{s''p^N}|_p.$$

But expanding

$$(1 + mp)^{s''p^N} = 1 + (s''p^N)mp + \frac{s''p^N(s''p^N - 1)}{2}(mp)^2 + \cdots + (mp)^{s''p^N}$$

shows that each term in $1 - (1 + mp)^{s''p^N}$ has at least p^{N+1}. Thus,

$$|n^s - n^{s'}|_p \le |p^{N+1}|_p = \frac{1}{p^{N+1}}.$$

In other words, if $s' - s$ is divisible by p^N, then $n^s - n^{s'}$ is divisible by p^{N+1}.

Thus, if $n \equiv 1 \pmod{p}$, it makes sense to define $f(s) = n^s$ for *any p-adic integer s* to be the *p*-adic integer which is the limit of n^{s_i} as s_i runs through any sequence of nonnegative integers which approach s (for example, the partial sums of the *p*-adic expansion of s). Then $f(s)$ is a continuous function from $\mathbb{Z}_p$ to $\mathbb{Z}_p$.

We can do a little better—allowing any n not divisible by *p*—if we're willing to insist that s and s' be congruent modulo $(p - 1)$, as well as modulo a high power of *p*. That is, we fix some $s_0 \in \{0, 1, 2, 3, \ldots, p - 2\}$, and, instead of considering n^s for all nonnegative integers s, we consider n^s for all nonnegative integers s congruent to our fixed s_0 modulo $(p - 1)$. Letting $s = s_0 + (p - 1)s_1$, we are looking at $n^{s_0 + (p-1)s_1}$ for s_1 any nonnegative integer. We can do this because then

$$n^s = n^{s_0}(n^{p-1})^{s_1},$$

and for every n not divisible by p we have $n^{p-1} \equiv 1 \pmod{p}$. Thus, we are in the situation of the last paragraph with n^{p-1} in place of n and s_1 in place of s (and a *constant* factor n^{s_0} thrown in).

Another way of expressing this function is as follows. Let S_{s_0} be the set of nonnegative integers congruent to s_0 mod $(p - 1)$. S_{s_0} is a dense subset of $\mathbb{Z}_p$ (Exercise 7 below). The function $f \cdot S_{s_0} \to \mathbb{Z}_p$ defined by $f(s) = n^s$ can be extended by continuity to a function $f: \mathbb{Z}_p \to \mathbb{Z}_p$. Notice that the function f depends on s_0 as well as on n. But when $p = 2$ we have $s_0 = 0$, and so if n is odd, then n^s is continuous as a function of all nonnegative integers.

If $n \equiv 0 \pmod{p}$, we are out of luck. This is because $n^{s_i} \to 0$ *p*-adically for *any* increasing sequence of nonnegative integers. And if $s \in \mathbb{Z}_p$ is not itself a nonnegative integer, *any* sequence of nonnegative integers which approach s *p*-adically must include arbitrarily large integers. It follows that the zero function is the only possible candidate for n^s, and that's absurd.

One final remark: the above discussion applies word-for-word to the function $1/n^s$ (Exercise 8 below).

Now let's look at the Riemann zeta-function

$$\zeta(s) = \sum_{n=1}^{\infty} \frac{1}{n^s} \quad (s > 1).$$

The naive way to try to interpolate $\zeta(s)$ *p*-adically would be to interpolate each term individually and then add the result. This won't work, because even the terms which can be interpolated—those for which $p \nmid n$—form an infinite sum which diverges in $\mathbb{Z}_p$. However, let's forget that for a moment and look at the terms one-by-one.

The first thing we'll want to do is get rid of the terms $1/n^s$ with n divisible by p. We do this as follows:

$$\zeta(s) = \sum_{n=1, p \nmid n}^{\infty} \frac{1}{n^s} + \sum_{n=1, p \mid n}^{\infty} \frac{1}{n^s} = \sum_{n=1, p \nmid n}^{\infty} \frac{1}{n^s} + \sum_{n=1}^{\infty} \frac{1}{p^s n^s}$$

$$= \sum_{n=1, p \nmid n}^{\infty} \frac{1}{n^s} + \frac{1}{p^s} \zeta(s);$$

$$\zeta(s) = \frac{1}{1 - (1/p^s)} \sum_{n=1, p \nmid n}^{\infty} \frac{1}{n^s}.$$

27

It is this last sum

$$\zeta^*(s) \underset{\text{def}}{=} \sum_{n=1, p\nmid n}^{\infty} \frac{1}{n^s} = \left(1 - \frac{1}{p^s}\right)\zeta(s)$$

which we will be dealing with later. This process is known as "taking out the *p*-Euler factor." The reason is that $\zeta(s)$ has a famous expansion (see Exercise 1 below)

$$\zeta(s) = \prod_{\text{primes } q} \frac{1}{1 - (1/q^s)}.$$

The factor $1/[1 - (1/q^s)]$ corresponding to the prime q is called the "*q*-Euler factor." Thus, multiplying $\zeta(s)$ by $[1 - (1/p^s)]$ amounts to removing the *p*-Euler factor:

$$\zeta^*(s) = \prod_{\text{primes } q \neq p} \frac{1}{1 - (1/q^s)}.$$

The second thing we'll want to do when interpolating $\zeta(s)$ is fix $s_0 \in \{0, 1, 2, \ldots, p - 2\}$ and only let s vary over nonnegative integers $s \in S_{s_0} = \{s \mid s \equiv s_0 \pmod{p - 1}\}$.

It will turn out that the numbers $(-B_{2k}/2k)$ arrived at in §1, when multiplied by $(1 - p^{2k-1})$, can be interpolated for $2k \in S_{2s_0}$ $(2s_0 \in \{0, 2, 4, \ldots, p - 3\})$. Note that we are not multiplying by $[1 - (1/p^{2k})]$, as you might expect, but rather by the Euler term with $2k$ replaced by $1 - 2k$: $1 - (1/p^{1-2k}) = 1 - p^{2k-1}$. The reason why this replacement $2k \leftrightarrow 1 - 2k$ is natural will be discussed in §7. (We'll see that the "interesting factor" $-B_{2k}/2k$ in $\zeta(2k)$ actually equals $\zeta(1 - 2k)$; $\zeta(x)$ and $\zeta(1 - x)$ are connected by a "functional equation.")

More precisely, we will show that, if $2k, 2k' \in S_{2k_0}$ (where $2k_0 \in \{2, 4, \ldots, p - 3\}$; there's a slight complication when $k_0 = 0$), and if $k \equiv k' \pmod{p^N}$, then (see §6)

$$(1 - p^{2k-1})(-B_{2k}/2k) \equiv (1 - p^{2k'-1})(-B_{2k'}/2k') \pmod{p^{N+1}}.$$

These congruences were first discovered by Kummer a century ago, but their interpretation in terms of *p*-adic interpolation of the Riemann ζ-function was only discovered in 1964 by Kubota and Leopoldt.

EXERCISES

1. Prove that

$$\zeta(s) = \prod_{\text{primes } q} \frac{1}{(1 - q^{-s})} \quad \text{for } s > 1.$$

2. Prove that

$$\prod_{r=1}^{k} \frac{(1 - \sin^2(\pi x/n)/\sin^2(\pi r/n))}{(1 - x^2/r^2)} \to 1 \quad \text{as } n = 2k + 1 \to \infty.$$

3. Use the relationship $e^{ix} = \cos x + i \sin x$ for e to a complex power to show that $\sinh x = -i \sin ix$. Give another argument for how the infinite product for $\sinh x$ follows from the infinite product for $\sin x$.

4. Prove that $B_k = 0$ if k is an odd number greater than 1.

5. Use the formula for $\zeta(2k)$, along with Stirling's asymptotic formula $n! \sim \sqrt{2\pi n}\, n^n e^{-n}$ (where $\sim$ means that the ratio of the two sides $\to 1$ as $n \to \infty$) to find an asymptotic estimate for the usual Archimedean absolute value of B_{2k}.

6. Use the discussion of n^s in §2 to compute the following through the p^4-place:
    (i) $11^{1/601}$ in $\mathbb{Q}_5$ (ii) $\sqrt{1/10}$ in $\mathbb{Q}_3$ (iii) $(-6)^{2 + 4 \cdot 7 + 3 \cdot 7^2 + 7^3 + \cdots}$ in $\mathbb{Q}_7$.

7. Prove that for any fixed $s_0 \in \{0, 1, \ldots, p-2\}$, the set of nonnegative integers congruent to s_0 modulo $(p-1)$ is dense in $\mathbb{Z}_p$, i.e., any number in $\mathbb{Z}_p$ can be approximated by such numbers.

8. What happens to the discussion in §2 if we take $n \in \mathbb{Z}_p$ instead of taking n to be a positive integer? What happens if we replace the function $f(s) = n^s$ by $f(s) = 1/n^s$? Note that this is the same as replacing "nonnegative integer" by "nonpositive integer" when defining the dense subset of $\mathbb{Z}_p$ from which we extend f.

9. Let χ be the function on the positive integers defined by:

$$\chi(n) = \begin{cases} 1, & \text{if } n \equiv 1 \pmod 4; \\ -1, & \text{if } n \equiv 3 \pmod 4; \\ 0, & \text{if } 2|n. \end{cases}$$

Define $L_\chi(s) \underset{\text{def}}{=} \sum_{n=1}^{\infty} (\chi(n)/n^s) = 1 - (1/3^s) + (1/5^s) - (1/7^s) + \cdots$. Prove that $L_\chi(s)$ converges absolutely if $s > 1$ and conditionally if $s > 0$. Find $L_\chi(1)$. Find an Euler product for $L_\chi(s)$ and for $L_\chi^*(s) \underset{\text{def}}{=} \sum_{n \geq 1, p \nmid n} (\chi(n)/n^s)$. (It turns out that there is a formula similar to Theorem 4 for $L_\chi(2k+1)$ (i.e., for positive *odd* rather than even integers) with B_n replaced by

$$B_{\chi,n} \underset{\text{def}}{=} n! \text{ times the coefficient of } t^n \text{ in } \frac{te^t}{e^{4t} - 1} - \frac{te^{3t}}{e^{4t} - 1} \left(= \frac{-t}{e^t + e^{-t}} \right).)$$

Note: Exercise 9 is a special case of the following situation. Let N be a positive integer. Let $(\mathbb{Z}/N\mathbb{Z})^\times$ be the multiplicative group of integers prime to N modulo N. Let $\chi : (\mathbb{Z}/N\mathbb{Z})^\times \to \mathbb{C}^\times$ be a group homomorphism from $(\mathbb{Z}/N\mathbb{Z})^\times$ to the multiplicative group of nonzero complex numbers. (It is easy to see that the image of χ can only contain roots of 1 in $\mathbb{C}$.) Suppose that χ is "primitive," which means that there is no M dividing N, $1 \leq M < N$, such that the value of χ on elements of $(\mathbb{Z}/N\mathbb{Z})^\times$ only depends on their value modulo M. Consider χ as a function on all positive integers n by letting $\chi(n)$ equal $\chi(n \text{ modulo } N)$ if n is prime to N and $\chi(n) = 0$ if n and N have a common factor greater than 1. χ is called a "character of conductor N."
    Now define

$$L_\chi(s) \underset{\text{def}}{=} \sum_{n=1}^{\infty} \frac{\chi(n)}{n^s}.$$

29

It can then be shown that for χ nontrivial a formula similar to Theorem 4 holds with B_n replaced by

$$B_{\chi,n} \underset{\text{def}}{=} n! \text{ times the coefficient of } t^n \text{ in } \sum_{a=1}^{N-1} \frac{\chi(a)te^{at}}{e^{Nt}-1}.$$

The formula gives L_χ of even integers if $\chi(-1) = 1$ and L_χ of odd integers if $\chi(-1) = -1$. (See Iwasawa, *Lectures on p-adic L-functions*.)

In addition, we have the formula

$$L_\chi(1) = \sum_{n=1}^{\infty} \frac{\chi(n)}{n} = \begin{cases} -\dfrac{\tau(\chi)}{N} \sum\limits_{a=1}^{N-1} \bar\chi(a) \log \sin \dfrac{a\pi}{N}, & \text{if } \chi(-1) = 1; \\[2ex] \dfrac{\pi i \tau(\chi)}{N^2} \sum\limits_{a=1}^{N-1} \bar\chi(a)\cdot a, & \text{if } \chi(-1) = -1, \end{cases}$$

where the bar over χ denotes the complex conjugate character: $\bar\chi(a) \underset{\text{def}}{=} \overline{\chi(a)}$, and where

$$\tau(\chi) \underset{\text{def}}{=} \sum_{a=1}^{N-1} \chi(a)e^{2\pi i a/N}$$

(this is known as a "Gauss sum"). (For a proof, see Borevich and Shafarevich, *Number Theory*, p. 332–336.)

10. Use the formula for $L_\chi(1)$ in the above note to check the value for $L_\chi(1)$ in Exercise 9 and to prove that:

(a) $\dfrac{1}{1} - \dfrac{1}{2} + \dfrac{1}{4} - \dfrac{1}{5} + \dfrac{1}{7} - \dfrac{1}{8} + \dfrac{1}{10} - \dfrac{1}{11} + \cdots + \dfrac{1}{3k+1} - \dfrac{1}{3k+2} + \cdots$

$= \dfrac{\pi}{3\sqrt{3}}$;

(b) $\dfrac{1}{1} - \dfrac{1}{3} - \dfrac{1}{5} + \dfrac{1}{7} + \dfrac{1}{9} - \dfrac{1}{11} - \dfrac{1}{13} + \dfrac{1}{15} + \dfrac{1}{17} - \dfrac{1}{19} - \dfrac{1}{21} + \dfrac{1}{23} + \cdots$

$= \dfrac{\log(1 + \sqrt{2})}{\sqrt{2}}$.

3. *p*-adic distributions

The metric space $\mathbb{Q}_p$ has a "basis of open sets" consisting of all sets of the form $a + p^N\mathbb{Z}_p = \{x \in \mathbb{Q}_p \mid |x - a|_p \leq (1/p^N)\}$ for $a \in \mathbb{Q}_p$ and $N \in \mathbb{Z}$. This means that any open subset of $\mathbb{Q}_p$ is a union of open subsets of this type. We shall sometimes abbreviate $a + p^N\mathbb{Z}_p$ as $a + (p^N)$, and in this chapter we shall call a set of this type an "interval" (in other contexts we often call such a set a "disc"). Notice that all intervals are closed as well as open, since the complement of $a + (p^N)$ is the union over all $a' \in \mathbb{Q}_p$ such that $a' \notin a + (p^N)$ of the open sets $a' + (p^N)$.

Recall that $\mathbb{Z}_p$ is sequentially compact: every sequence of *p*-adic integers has a convergent subsequence (see Exercise 19 of §I.5). The same is easily seen to be true for any interval or finite union of intervals. In a metric space X, the property of sequential compactness of a set $S \subset X$ is equivalent to the following property, called "compactness": every time S is contained in a union of sets, it is contained in a union of finitely many of those open sets ("every open covering has a finite subcovering"). (See Simmons, *Introduction to Topology and Modern Analysis*, §24, for this equivalence; this book is also a good standard reference for other concepts from general topology.) It then follows (see Exercise 1 below) that an open subset of $\mathbb{Q}_p$ is compact if and only if it is a finite union of intervals. It is this type of open set, which we call a "compact-open," that repeatedly occurs in this section.

Definition. Let X and Y be two topological spaces. A map $f: X \to Y$ is called *locally constant* if every point $x \in X$ has a neighborhood U such that $f(U)$ is a single element of Y.

It is trivial to see that a locally constant function is continuous.

The concept of a locally constant function is not very useful in classical situations, because there usually aren't any, except for constants. This is the case whenever X is connected, for example $\mathbb{R}$ or $\mathbb{C}$.

But for us X will be a compact-open subset of $\mathbb{Q}_p$ (usually $\mathbb{Z}_p$ or $\mathbb{Z}_p^\times = \{x \in \mathbb{Z}_p \mid |x|_p = 1\}$). Then X has many nontrivial locally constant functions. In fact, $f: X \to \mathbb{Q}_p$ is locally constant precisely when f is a finite linear combination of characteristic functions of compact-open sets (see Exercise 4 below).

Locally constant functions play the same role for p-adic X that step-functions play when $X = \mathbb{R}$ in defining integrals by means of Riemann sums.

Now let X be a compact-open subset of $\mathbb{Q}_p$, such as $\mathbb{Z}_p$ or $\mathbb{Z}_p^\times$.

Definition. A *p-adic distribution* μ on X is a $\mathbb{Q}_p$-linear vector space homomorphism from the $\mathbb{Q}_p$-vector space of locally constant functions on X to $\mathbb{Q}_p$. If $f: X \to \mathbb{Q}_p$ is locally constant, instead of writing $\mu(f)$ for the value of μ at f, we usually write $\int f\mu$.

Equivalent definition (see Exercise 4 below). A *p*-adic distribution μ on X is an additive map from the set of compact-opens in X to $\mathbb{Q}_p$; this means that if $U \subset X$ is the disjoint union of compact-open sets $U_1, U_2, \ldots, U_n$, then

$$\mu(U) = \mu(U_1) + \mu(U_2) + \cdots + \mu(U_n).$$

By "equivalent definition," we mean that any μ in the second sense "extends" uniquely to a μ in the first sense, and any μ in the first sense "restricts" to a μ in the second sense. More precisely, if we have a distribution μ in the sense of the first definition, we get a distribution (also denoted μ) in the sense of the second definition by letting

$$\mu(U) = \int (\text{characteristic function of } U)\mu,$$

for every compact-open U. If we have a distribution μ in the sense of the second definition, we get a distribution in the sense of the first definition by first letting

$$\int (\text{characteristic function of } U)\mu = \mu(U),$$

and then defining $\int f\mu$ for locally constant f by writing f as a linear combination of characteristic functions.

Proposition. *Every map μ from the set of intervals contained in X to $\mathbb{Q}_p$ for which*

$$\mu(a + (p^N)) = \sum_{b=0}^{p-1} \mu(a + bp^N + (p^{N+1}))$$

whenever $a + (p^N) \subset X$, extends uniquely to a p-adic distribution on X.

PROOF. Every compact-open $U \subset X$ can be written as a finite disjoint union of intervals: $U = \bigcup I_i$ (see Exercise 1). We then define $\mu(U) \underset{\text{def}}{=} \sum \mu(I_i)$. (This is the only possible value of $\mu(U)$ if μ is to be additive.) To check that $\mu(U)$ does not depend on the partitioning of U into intervals, we first note that any two partitions $U = \bigcup I_i$ and $U = \bigcup I_i'$ of U into a disjoint union of intervals have a common subpartition ("finer" than both) which is of the form $I_i = \bigcup_j I_{ij}$, where, if $I_i = a + (p^N)$, then the I_{ij}'s run through all intervals $a' + (p^{N'})$ for some fixed $N' > N$ and for variable a' which are $\equiv a \pmod{p^N}$. Then, by repeated application of the equality in the statement of the proposition, we have:

$$\mu(I_i) = \mu(a + (p^N)) = \sum_{j=0}^{p^{N'-N}-1} \mu(a + jp^N + (p^{N'})) = \sum_j \mu(I_{ij}).$$

Hence $\sum_i \mu(I_i) = \sum_{i,j} \mu(I_{ij})$. Thus, $\sum_i \mu(I_i) = \sum_i \mu(I_i')$, because both sides equal the sum over the common subpartition. It is now clear that μ is additive. Namely, if U is a disjoint union of U_i, we write each U_i as a disjoint union of intervals I_{ij}, so that $U = \bigcup_{i,j} I_{ij}$, and

$$\mu(U) = \sum_{i,j} \mu(I_{ij}) = \sum_i \sum_j \mu(I_{ij}) = \sum_i \mu(U_i). \qquad \square$$

We now give some simple examples of *p*-adic distributions.

(1) The Haar distribution μ_{Haar}. Define

$$\mu_{\text{Haar}}(a + (p^N)) \underset{\text{def}}{=} \frac{1}{p^N}.$$

This extends to a distribution on $\mathbb{Z}_p$ by the proposition, since

$$\sum_{b=0}^{p-1} \mu_{\text{Haar}}(a + bp^N + (p^{N+1})) = \sum_{b=0}^{p-1} \frac{1}{p^{N+1}} = \frac{1}{p^N}$$

$$= \mu_{\text{Haar}}(a + (p^N)).$$

This is the unique distribution (up to a constant multiple) which is "translation invariant," meaning that for all $a \in \mathbb{Z}_p$ we have $\mu_{\text{Haar}}(a + U) = \mu_{\text{Haar}}(U)$, where $a + U \underset{\text{def}}{=} \{x \in \mathbb{Z}_p \mid x - a \in U\}$.

(2) The Dirac distribution μ_α concentrated at $\alpha \in \mathbb{Z}_p$ (α is fixed). Define

$$\mu_\alpha(U) \underset{\text{def}}{=} \begin{cases} 1, & \text{if } \alpha \in U; \\ 0, & \text{otherwise.} \end{cases}$$

It is trivial to check that μ_α is additive. Note that $\int f \mu_\alpha = f(\alpha)$ for locally constant f.

(3) The Mazur distribution μ_{Mazur}. First, without loss of generality, when we write $a + (p^N)$ we may assume that a is a rational integer between 0 and $p^N - 1$. Assuming this, we define

$$\mu_{\text{Mazur}}(a + (p^N)) \underset{\text{def}}{=} \frac{a}{p^N} - \frac{1}{2}.$$

We postpone the verification that μ_{Mazur} has the additivity property in the proposition, since this will come as a special case of a more general result in the next section.

Notice one important difference between the distributions μ_{Haar} and μ_{Mazur} and classical measures. In these two p-adic examples, as the interval being measured "shrinks" (i.e., as $N \to \infty$), its measure in terms of μ *increases* as a number in $\mathbb{Q}_p$, namely:

$$|\mu_{\text{Haar}}(a + (p^N))|_p = \left|\frac{1}{p^N}\right|_p = p^N;$$

and, if $p \nmid a$ (and if $N > 1$ in the case $p = 2$), then

$$|\mu_{\text{Mazur}}(a + (p^N))|_p = \left|\frac{a}{p^N} - \frac{1}{2}\right|_p = p^N.$$

We'll deal with this peculiarity later.

EXERCISES

1. Give a direct proof that $\mathbb{Z}_p$ is compact (i.e., that any open covering of $\mathbb{Z}_p$ has a finite subcovering). Then prove that an open set in $\mathbb{Z}_p$ is compact if and only if it can be written as a finite disjoint union of intervals. Note that any interval can be written as a disjoint union of p "equally long" subintervals: $a + (p^n) = \bigcup_{b=0}^{p-1} a + bp^n + (p^{n+1})$. Prove that any partition of an interval into a disjoint union of subintervals can be obtained by applying this process a finite number of times.

2. Give an example of a noncompact open subset of $\mathbb{Z}_p$.

3. Let U be an open subset of a topological space X. Show that the characteristic function $f: X \to \mathbb{Z}$ defined by

$$f(x) = \begin{cases} 1, & \text{if } x \in U; \\ 0, & \text{otherwise,} \end{cases}$$

is locally constant if $X = \mathbb{Z}_p$ and U is a compact-open but is not locally constant for any open set U if $X = \mathbb{R}$ (unless U is $\mathbb{R}$ itself or the empty set).

4. Let X be a compact-open subset of $\mathbb{Q}_p$. Show that $f: X \to \mathbb{Q}_p$ is locally constant if and only if it is a finite linear combination with coefficients in $\mathbb{Q}_p$ of characteristic functions of compact-opens in X. Then prove that the two definitions of a distribution on X are equivalent.

5. If $\alpha \in \mathbb{Q}_p$, $|\alpha|_p = 1$, show that $\mu_{\text{Haar}}(\alpha U) = \mu_{\text{Haar}}(U)$ for all compact-open U, where αU denotes $\{\alpha x \,|\, x \in U\}$.

6. Let $f: \mathbb{Z}_p \to \mathbb{Q}_p$ be the locally constant function defined by: $f(x) =$ the first digit in the p-adic expansion of x. Find $\int f\mu$ when: (1) $\mu =$ the Dirac distribution μ_a; (2) $\mu = \mu_{\text{Haar}}$; (3) $\mu = \mu_{\text{Mazur}}$.

7. Let μ be the function of intervals $a + (p^N)$ which is defined as follows ([ ] = greatest integer function):

$$\mu(a + (p^N)) = \begin{cases} p^{-[(N+1)/2]}, & \text{if the first } [N/2] \text{ digits in } a \text{ corresponding to odd} \\ & \text{powers of } p \text{ vanish;} \\ 0, & \text{otherwise.} \end{cases}$$

Prove that μ extends to a distribution on $\mathbb{Z}_p$.

8. Discuss how one could go about making up examples of p-adic distributions μ on $\mathbb{Z}_p$ with various growth rates (i.e., rates of growth of $\max_{0 \le a < p^N} |\mu(a + (p^N))|_p$ as N increases).

4. Bernoulli distributions

We first define the Bernoulli polynomials $B_k(x)$. Consider the function in two variables t and x

$$\frac{te^{xt}}{e^t - 1} = \left(\sum_{k=0}^{\infty} B_k \frac{t^k}{k!} \right) \left(\sum_{k=0}^{\infty} \frac{(xt)^k}{k!} \right).$$

In this product, we collect the terms with t^k, obtaining for each k a polynomial in x, and we define $B_k(x)$ to be $k!$ times this polynomial:

$$\frac{te^{xt}}{e^t - 1} = \sum_{k=0}^{\infty} B_k(x) \frac{t^k}{k!}.$$

The first few Bernoulli polynomials are:

$$B_0(x) = 1, \quad B_1(x) = x - \tfrac{1}{2}, \quad B_2(x) = x^2 - x + \tfrac{1}{6},$$
$$B_3(x) = x^3 - \tfrac{3}{2}x^2 + \tfrac{1}{2}x, \dots .$$

Throughout this section, when we write $a + (p^N)$ we will assume that $0 \le a \le p^N - 1$. Fix a nonnegative integer k. We define a map $\mu_{B,k}$ on intervals $a + (p^N)$ by

$$\mu_{B,k}(a + (p^N)) = p^{N(k-1)} B_k\left(\frac{a}{p^N}\right).$$

Proposition. $\mu_{B,k}$ *extends to a distribution on* $\mathbb{Z}_p$ (called the "kth Bernoulli distribution").

PROOF. By the proposition in §3, we must show that

$$\mu_{B,k}(a + (p^N)) = \sum_{b=0}^{p-1} \mu_{B,k}(a + bp^N + (p^{N+1})).$$

The right-hand side equals

$$p^{(N+1)(k-1)} \sum_{b=0}^{p-1} B_k\left(\frac{a + bp^N}{p^{N+1}}\right),$$

so, multiplying by $p^{-N(k-1)}$ and setting $\alpha = a/p^{N+1}$, we must show that

$$B_k(p\alpha) = p^{k-1} \sum_{b=0}^{p-1} B_k\left(\alpha + \frac{b}{p}\right).$$

The right-hand side is, by the definition of $B_k(x)$, equal to $k!$ times the coefficient of t^k in

$$p^{k-1} \sum_{b=0}^{p-1} \frac{te^{(\alpha + b/p)t}}{e^t - 1} = \frac{p^{k-1}te^{\alpha t}}{e^t - 1} \sum_{b=0}^{p-1} e^{bt/p} = \frac{p^{k-1}te^{\alpha t}}{e^t - 1} \cdot \frac{e^t - 1}{e^{t/p} - 1},$$

by summing the geometric progression $\sum_{b=0}^{p-1} e^{bt/p}$. This expression equals

$$\frac{p^k(t/p)e^{(p\alpha)t/p}}{e^{t/p} - 1} = p^k \sum_{j=0}^{\infty} B_j(p\alpha) \frac{(t/p)^j}{j!},$$

again by the definition of $B_j(x)$. Hence, $k!$ times the coefficient of t^k is simply

$$p^k B_k(p\alpha)\left(\frac{1}{p}\right)^k = B_k(p\alpha),$$

as desired. $\qquad\square$

The first few $B_k(x)$ give us the following distributions:

$$\mu_{B,0}(a + (p^N)) = p^{-N}, \quad \text{i.e., } \mu_{B,0} = \mu_{\text{Haar}};$$

$$\mu_{B,1}(a + (p^N)) = B_1\left(\frac{a}{p^N}\right) = \frac{a}{p^N} - \frac{1}{2}, \quad \text{i.e., } \mu_{B,1} = \mu_{\text{Mazur}};$$

$$\mu_{B,2}(a + (p^N)) = p^N\left(\frac{a^2}{p^{2N}} - \frac{a}{p^N} + \frac{1}{6}\right),$$

and so on.

It can be shown that the Bernoulli polynomials are the only polynomials (up to a constant multiple) that can be used to define distributions in this way. We shall not need this fact, and so will not prove it. But it should be noticed that the Bernoulli polynomials $B_k(x)$ have appeared in an important and unique role in *p*-adic integration. This will turn out to be related to the appearance of the Bernoulli numbers B_k (which are the constant terms in the $B_k(x)$; see Exercise 1 below) in the formula for $\zeta(2k)$.

5. Measures and integration

Definition. A *p*-adic distribution μ on X is a *measure* if its values on compact-open $U \subset X$ are bounded by some constant $B \in \mathbb{R}$, i.e.,

$$|\mu(U)|_p \leq B \quad \text{for all compact-open } U \subset X.$$

The Dirac distribution μ_α for fixed $\alpha \in \mathbb{Z}_p$ is a measure, but none of the Bernoulli distributions are measures. There is a standard method, called "regularization," for turning Bernoulli distributions into measures. We first introduce some notation. If $\alpha \in \mathbb{Z}_p$, we let $\{\alpha\}_N$ be the rational integer between 0 and $p^N - 1$ which is $\equiv \alpha \pmod{p^N}$. If μ is a distribution and $\alpha \in \mathbb{Q}_p$, we let $\alpha\mu$ denote the distribution whose value on any compact-open is α times the value of μ: $(\alpha\mu)(U) = \alpha \cdot (\mu(U))$. Finally, if $U \subset \mathbb{Q}_p$ is a compact-open set and $\alpha \in \mathbb{Q}_p$, $\alpha \neq 0$, we let $\alpha U \underset{\text{def}}{=} \{x \in \mathbb{Q}_p \mid x/\alpha \in U\}$. It is trivial to check that the sum of two distributions (or measures) is a distribution (resp. measure), any scalar multiple $\alpha\mu$ of a distribution (or measure) μ is a distribution (resp. measure), and, if $\alpha \in \mathbb{Z}_p^\times$ and if μ is a distribution (or measure) on $\mathbb{Z}_p$, then the function μ' defined by $\mu'(U) = \mu(\alpha U)$ is a distribution (resp. measure) on $\mathbb{Z}_p$.

Now let α be any rational integer not equal to 1 and not divisible by p. Let $\mu_{B,k,\alpha}$—or, more briefly, $\mu_{k,\alpha}$—be the "regularized" Bernoulli distribution on $\mathbb{Z}_p$ defined by

$$\mu_{k,\alpha}(U) \underset{\text{def}}{=} \mu_{B,k}(U) - \alpha^{-k}\mu_{B,k}(\alpha U).$$

We will soon show that $\mu_{k,\alpha}$ is a measure. In any case, it's clearly a distribution by the remarks in the last paragraph.

We easily compute an explicit formula when $k = 0$ or 1. For $k = 0$, $\mu_{B,0} = \mu_{\text{Haar}}$, and it is easy to see that $\mu_{0,\alpha}(U) = 0$ for all U (see Exercise 5 of §3). If $k = 1$, we have

$$\mu_{1,\alpha}(a + (p^N)) = \frac{a}{p^N} - \frac{1}{2} - \frac{1}{\alpha}\left(\frac{\{\alpha a\}_N}{p^N} - \frac{1}{2}\right)$$

$$= \frac{(1/\alpha) - 1}{2} + \frac{a}{p^N} - \frac{1}{\alpha}\left(\frac{\alpha a}{p^N} - \left[\frac{\alpha a}{p^N}\right]\right)$$

(where [ ] means the greatest integer function)

$$= \frac{1}{\alpha}\left[\frac{\alpha a}{p^N}\right] + \frac{(1/\alpha) - 1}{2}.$$

Proposition. $|\mu_{1,\alpha}(U)|_p \leq 1$ *for all compact-open* $U \subset \mathbb{Z}_p$.

PROOF. Notice that $(\alpha^{-1} - 1)/2 \in \mathbb{Z}_p$, since $1/\alpha \in \mathbb{Z}_p$ and $1/2 \in \mathbb{Z}_p$ unless $p = 2$. If $p = 2$, then $\alpha^{-1} - 1 \equiv 0 \pmod 2$, and we're still OK. Since $[\alpha a/p^N] \in \mathbb{Z}$, it follows by the above formula that $\mu_{1,\alpha}(a + (p^N)) \in \mathbb{Z}_p$. Then, since every compact-open U is a finite disjoint union of intervals I_i, we may conclude that $|\mu_{1,\alpha}(U)|_p \leq \max |\mu_{1,\alpha}(I_i)|_p \leq 1$. $\qquad\qquad\qquad\qquad\qquad\square$

Thus, $\mu_{1,\alpha}$ is a measure—the first interesting example of a p-adic measure that we've come across. In fact, we'll soon see that $\mu_{1,\alpha}$ plays a fundamental role in p-adic integration, almost as fundamental as the role played by "dx" in real integration.

We next prove a key congruence that relates $\mu_{k,\alpha}$ to $\mu_{1,\alpha}$. The proof of this congruence at first looks unpleasantly computational, but it becomes more transparent if we think of an analogous situation in real calculus. Suppose that in taking integrals such as $\int f(\sqrt[k]{x})dx$ we want to make the change of variables $x \mapsto x^k$, i.e., to evaluate $\int f(x)d(x^k)$. The simple rule is: $d(x^k)/dx = kx^{k-1}$. Actually, $d(x^k)$ can be thought of as a "measure" μ_k on the real number line, which is defined by letting $\mu_k([a, b]) = b^k - a^k$; then μ_1 is the usual concept of length. The relation $d(x^k)/dx = kx^{k-1}$ actually means

$$\lim_{b \to a} \frac{\mu_k([a, b])}{\mu_1([a, b])} = ka^{k-1}.$$

Thus, in the Riemann sums $\sum f(x_i)\mu_k(I_i)$ in the limit as the I_i's all become smaller we may replace $\mu_k(I_i)$ by $kx^{k-1}\mu_1(I_i)$ and get $\int f(x)kx^{k-1}\,dx$.

The actual proof that $\lim_{b \to a} \mu_k([a, b])/\mu_1([a, b]) = ka^{k-1}$ uses the binomial expansion for $(a + h)^k$ (where $h = b - a$)—actually, only the first two terms $a^k + kha^{k-1}$ really matter. Similarly, in the p-adic case, when we show that $\mu_{k,\alpha}(I) \sim ka^{k-1}\mu_{1,\alpha}(I)$ if I is a small interval containing a, we also use the binomial expansion. Thus, Theorem 5 should be thought of as analogous to the theorem that $(d/dx)(x^k) = kx^{k-1}$ from real calculus. (Forget about the d_k on both sides of the congruence in Theorem 5; all it means is that, when we divide both sides by d_k, we must replace p^N by $p^{N - \text{ord}_p d_k}$, where $\text{ord}_p d_k$ is just a *constant* which doesn't matter for large N.)

Theorem 5. *Let d_k be the least common denominator of the coefficients of $B_k(x)$. Thus: $d_1 = 2$, $d_2 = 6$, $d_3 = 2$, etc. Then*

$$d_k\mu_{k,\alpha}(a + (p^N)) \equiv d_k ka^{k-1}\mu_{1,\alpha}(a + (p^N)) \pmod{p^N},$$

where both sides of this congruence lie in $\mathbb{Z}_p$.

PROOF. By Exercise 1 below, the polynomial $B_k(x)$ starts out

$$B_0 x^k + kB_1 x^{k-1} + \cdots = x^k - \frac{k}{2}x^{k-1} + \cdots.$$

Now

$$d_k\mu_{k,\alpha}(a + (p^N)) = d_k p^{N(k-1)}\left(B_k\left(\frac{a}{p^N}\right) - \alpha^{-k}B_k\left(\frac{\{\alpha a\}_N}{p^N}\right)\right).$$

The polynomial $d_k B_k(x)$ has integral coefficients and degree k. Hence we need only consider the leading two terms $d_k x^k - d_k(k/2)x^{k-1}$ of $d_k B_k(x)$, since our x has denominator p^N, so that the denominators in the lower terms of $d_k B_k(x)$ will be canceled by $p^{N(k-1)}$ with at least p^N left over. We also note that

$$\alpha a \equiv \{\alpha a\}_N \ (\mathrm{mod} \ p^N),$$

and

$$\frac{\{\alpha a\}_N}{p^N} = \frac{\alpha a}{p^N} - \left[\frac{\alpha a}{p^N}\right] \quad ([\ ] = \text{greatest integer function}).$$

Hence

$$d_k \mu_{k,\alpha}(a + (p^N)) \equiv d_k p^{N(k-1)}\left(\frac{a^k}{p^{Nk}} - \alpha^{-k}\left(\frac{\{\alpha a\}_N}{p^N}\right)^k\right.$$

$$\left. - \frac{k}{2}\left(\frac{a^{k-1}}{p^{N(k-1)}} - \alpha^{-k}\left(\frac{\{\alpha a\}_N}{p^N}\right)^{k-1}\right)\right) \quad (\mathrm{mod} \ p^N)$$

$$= d_k\left(\frac{a^k}{p^N} - \alpha^{-k}p^{N(k-1)}\left(\frac{\alpha a}{p^N} - \left[\frac{\alpha a}{p^N}\right]\right)^k\right.$$

$$\left. - \frac{k}{2}\left(a^{k-1} - \alpha^{-k}p^{N(k-1)}\left(\frac{\alpha a}{p^N} - \left[\frac{\alpha a}{p^N}\right]\right)^{k-1}\right)\right)$$

$$\equiv d_k\left(\frac{a^k}{p^N} - \alpha^{-k}\left(\frac{\alpha^k a^k}{p^N} - k\alpha^{k-1}a^{k-1}\left[\frac{\alpha a}{p^N}\right]\right)\right.$$

$$\left. - \frac{k}{2}(a^{k-1} - \alpha^{-k}(\alpha^{k-1}a^{k-1}))\right) \quad (\mathrm{mod} \ p^N)$$

$$= d_k k a^{k-1}\left(\frac{1}{\alpha}\left[\frac{\alpha a}{p^N}\right] + \frac{1/\alpha - 1}{2}\right)$$

$$= d_k k a^{k-1} \mu_{1,\alpha}(a + (p^N)). \qquad \square$$

Corollary. *$\mu_{k,\alpha}$ is a measure for all $k = 1, 2, 3, \ldots$ and any $\alpha \in \mathbb{Z}, \alpha \notin p\mathbb{Z}, \alpha \neq 1$.*

PROOF. We must show that $\mu_{k,\alpha}(a + (p^N))$ is bounded. But by Theorem 5,

$$|\mu_{k,\alpha}(a + (p^N))|_p \leq \max\left(\left|\frac{p^N}{d_k}\right|_p, |ka^{k-1}\mu_{1,\alpha}(a + (p^N))|_p\right)$$

$$\leq \max\left(\left|\frac{1}{d_k}\right|_p, |\mu_{1,\alpha}(a + (p^N))|_p\right).$$

But $|\mu_{1,\alpha}(a + (p^N))|_p \leq 1$, and d_k is fixed. $\qquad \square$

What is the purpose of going to all this fuss to modify ("regularize") Bernoulli distributions to get measures? The answer is that for an unbounded distribution μ, $\int f\mu$ is defined by definition as long as f is locally constant, but you run into problems if you try to use limits of Riemann sums to extend integration to continuous functions f.

For example, let $\mu = \mu_{\text{Haar}}$, and take the simple function $f: \mathbb{Z}_p \to \mathbb{Z}_p$ given by $f(x) = x$. Let's form the Riemann sums. Given a function f, for any N we divide up $\mathbb{Z}_p$ into $\bigcup_{a=0}^{p^N-1} (a + (p^N))$, we let $x_{a,N}$ be an arbitrary point in the ath interval, and we define the Nth Riemann sum of f corresponding to $\{x_{a,N}\}$ as

$$S_{N,\{x_{a,N}\}}(f) \underset{\text{def}}{=} \sum_{a=0}^{p^N-1} f(x_{a,N})\mu(a + (p^N)).$$

In our example, this sum equals

$$\sum_{a=0}^{p^N-1} x_{a,N} \frac{1}{p^N}.$$

For example, if we simply choose $x_{a,N} = a$, we obtain

$$p^{-N} \sum_{a=0}^{p^N-1} a = p^{-N} \frac{(p^N - 1)(p^N)}{2} = \frac{p^N - 1}{2}.$$

This sum has a limit in $\mathbb{Q}_p$ as $N \to \infty$, namely $-1/2$. But if, instead of $x_{a,N} = a \in a + (p^N)$, we change one of the $x_{a,N}$ to $a + a_0 p^N \in a + (p^N)$ for each N, where a_0 is some fixed p-adic integer, we then obtain

$$p^{-N}\left(\sum_{a=0}^{p^N-1} a + a_0 p^N\right) = \frac{p^N - 1}{2} + a_0,$$

whose limit is $a_0 - \frac{1}{2}$. Thus, the Riemann sums do not have a limit which is independent of the choice of points in the intervals.

A "measure" μ is not much good, and has no right to be called a measure, if you can't integrate continuous functions with respect to it. (This is a slight exaggeration—see Exercises 8–10 below.) Now we show that bounded distributions earn their name of "measure".

Recall that X is a compact-open subset of $\mathbb{Q}_p$, such as $\mathbb{Z}_p$ or $\mathbb{Z}_p^\times$. (For simplicity, let $X \subset \mathbb{Z}_p$.)

Theorem 6. *Let μ be a p-adic measure on X, and let $f: X \to \mathbb{Q}_p$ be a continuous function. Then the Riemann sums*

$$S_{N,\{x_{a,N}\}} \underset{\text{def}}{=} \sum_{\substack{0 \leq a < p^N \\ a + (p^N) \subset X}} f(x_{a,N})\mu(a + (p^N))$$

(where the sum is taken over all a for which $a + (p^N) \subset X$, and $x_{a,N}$ is chosen in $a + (p^N)$) converge to a limit in $\mathbb{Q}_p$ as $N \to \infty$ which does not depend on the choice of $\{x_{a,N}\}$.

PROOF. Suppose that $|\mu(U)|_p \leq B$ for all compact-open $U \subset X$. We first estimate for $M > N$

$$|S_{N,\{x_{a,N}\}} - S_{M,\{x_{a,M}\}}|_p.$$

By writing X as a finite union of intervals, we can choose N large enough so that every $a + (p^N)$ is either $\subset X$ or disjoint from X. We rewrite $S_{N,\{x_{a,N}\}}$ as follows, using the additivity of μ:

$$\sum_{\substack{0 \le a < p^M \\ a + (p^M) \subset X}} f(x_{a,N})\mu(a + (p^M))$$

(where $\bar{a}$ denotes the least nonnegative residue of $a \bmod p^N$). We further assume that N is large enough so that $|f(x) - f(y)|_p < \varepsilon$ whenever $x \equiv y$ (mod p^N). (Note that continuity implies *uniform* continuity, since X is *compact*; this is an easy exercise, or else see Simmons' book.) Then

$$|S_{N,\{x_{a,N}\}} - S_{M,\{x_{a,M}\}}|_p = \left| \sum_{\substack{0 \le a < p^M \\ a + (p^M) \subset X}} (f(x_{\bar{a},N}) - f(x_{a,M}))\mu(a + (p^M)) \right|_p$$

$$\le \max(|f(x_{\bar{a},N}) - f(x_{a,M})|_p \cdot |\mu(a + (p^M))|_p)$$

$$\le \varepsilon B,$$

since $x_{\bar{a},N} \equiv x_{a,M}$ (mod p^N). Since ε is arbitrary and B is fixed, the Riemann sums have a limit.

It follows similarly that this limit is independent of $\{x_{a,N}\}$. Namely,

$$|S_{N,\{x_{a,N}\}} - S_{N,\{x'_{a,N}\}}|_p = \left| \sum_{\substack{0 \le a < p^N \\ a + (p^N) \subset X}} (f(x_{a,N}) - f(x'_{a,N}))\mu(a + (p^N)) \right|_p$$

$$\le \max_a(|f(x_{a,N}) - f(x'_{a,N})|_p \cdot |\mu(a + (p^N))|_p)$$

$$\le \varepsilon B. \qquad \square$$

Definition. If $f: X \to \mathbb{Q}_p$ is a continuous function and μ is a measure on X, we define $\int f\mu$ to be the limit of the Riemann sums, the existence of which was just proved. (Note that, if f is locally constant, this definition agrees with the earlier meaning of $\int f\mu$.)

The following simple but important facts follow immediately from this definition.

Proposition. *If* $f: X \to \mathbb{Q}_p$ *is a continuous function such that* $|f(x)|_p \le A$ *for all* $x \in X$, *and if* $|\mu(U)|_p \le B$ *for all compact-open* $U \subset X$, *then*

$$\left| \int f\mu \right|_p \le A \cdot B.$$

Corollary. *If* $f, g: X \to \mathbb{Q}_p$ *are two continuous functions such that* $|f(x) - g(x)|_p \le \varepsilon$ *for all* $x \in X$, *and if* $|\mu(U)|_p \le B$ *for all compact-open* $U \subset X$, *then*

$$\left| \int f\mu - \int g\mu \right|_p \le \varepsilon B.$$

EXERCISES

1. Show that $B_k(x) = \sum_{i=0}^{k} \binom{k}{i} B_i x^{k-i}$, and, in particular, $B_k(0) = B_k$. Further show that

$$\int_0^1 B_k(x)\, dx = \begin{cases} 1 & \text{if } k = 0, \\ 0 & \text{otherwise,} \end{cases} \quad \text{and that} \quad \frac{d}{dx} B_k(x) = k B_{k-1}(x).$$

2. Prove that no distribution μ (except for the identically zero distribution) has the property that

$$\max_{0 \le a < p^N} |\mu(a + (p^N))|_p \to 0 \quad \text{as } N \to \infty.$$

3. What is $\mu_{B,k}(\mathbb{Z}_p)$? $\mu_{B,k}(p\mathbb{Z}_p)$? $\mu_{B,k}(\mathbb{Z}_p^\times)$?

4. Prove that a p-adic distribution μ is a measure if and only if for some nonzero $a \in \mathbb{Z}_p$ the distribution $a \cdot \mu$ takes values in $\mathbb{Z}_p$. Prove that the set of measures on X is a $\mathbb{Q}_p$-vector space.

5. Express $\mu_{k,\alpha}(\mathbb{Z}_p)$ and $\mu_{k,\alpha}(\mathbb{Z}_p^\times)$ in terms of α and k. Find $\int_{\mathbb{Z}_p^\times} f \mu_{1,\alpha}$ if $f(x) = \sum_{i=0}^{n} a_i x^i$.

6. Let p be an odd prime. For any $a = 0, 1, \ldots, p^n - 1$, let S_a denote the sum of the p-adic digits in a. Show that $\mu(a + (p^n)) = (-1)^{S_a}$ gives a measure on $\mathbb{Z}_p$, and that $\int_{\mathbb{Z}_p} f\mu = f(0)$ for any continuous even function f (i.e., for which $f(-x) = f(x)$).

7. Let $p > 2$, $f(x) = 1/x$, and $\alpha = 1 + p$. Prove that $\int_{\mathbb{Z}_p^\times} f \mu_{1,\alpha} \equiv -1 \pmod{p}$. If $p = 2$ let $\alpha = 5$, and prove that $\int_{\mathbb{Z}_2^\times} f \mu_{1,\alpha} \equiv 2 \pmod 4$.

8. A distribution μ on X is called "boundedly increasing" if $\max_{0 \le a < p^N} |p^N \mu(a + (p^N))|_p \to 0$ as $N \to \infty$, i.e., μ "increases strictly slower than μ_{Haar}." Prove that Theorem 6 holds for μ if we assume that $f: X \to \mathbb{Q}_p$ satisfies the Lipschitz condition: there exists an $A \in \mathbb{R}$ such that for all $x, y \in X$

$$|f(x) - f(y)|_p \le A|x - y|_p.$$

(This concept was introduced by Manin and applied by him to p-adic interpolation of certain Hecke series.)

9. Let μ be the distribution defined in Exercise 7 of §3. Check that μ is boundedly increasing. Let $f: \mathbb{Z}_p \to \mathbb{Z}_p$ be the function $f(x) = x$. Evaluate $\int f\mu$, which we know is well defined by the previous problem.

10. Let r be a positive real number. A function $f: \mathbb{Z}_p \to \mathbb{Q}_p$ is called (by Mazur) "of type r" if there exists $A \in \mathbb{R}$ such that for all $x, x' \in \mathbb{Z}_p$ we have

$$|f(x) - f(x')|_p \le A|x - x'|_p^r.$$

Note that any such function is continuous. If $r \ge 1$, then f is Lipschitz (see Exercise 8). Now let μ be a p-adic distribution on $\mathbb{Z}_p$ such that for some positive $s \in \mathbb{R}$

$$p^{-Ns} \max_{0 \le a < p^N} |\mu(a + (p^N))|_p \to 0 \quad \text{as } N \to \infty.$$

Prove the analogue of Theorem 6 for such a μ and functions of type r whenever $r \ge s$.

6. The *p*-adic ζ-function as a Mellin–Mazur transform

If X is a compact-open subset of $\mathbb{Z}_p$, any measure μ on $\mathbb{Z}_p$ can be restricted to X. This means we define a measure μ^* on X by setting $\mu^*(U) = \mu(U)$ whenever U is a compact-open in X. In terms of integrating functions, we have

$$\int f\mu^* = \int f\cdot(\text{characteristic function of } X)\mu.$$

We shall use the notation $\int_X f\mu$ for this restricted integral $\int f\mu^*$.

We said that what we want to interpolate is $-B_k/k$. We have the simple relationship

$$\int_{\mathbb{Z}_p} 1\cdot\mu_{B,k} = \mu_{B,k}(\mathbb{Z}_p) = B_k$$

(see Exercise 3 of §5). Hence we want to interpolate the numbers $-(1/k)\int_{\mathbb{Z}_p} 1\mu_{B,k}$.

For different k are the distributions $\mu_{B,k}$ related to each other in any straightforward way? Not quite, but the *regularized* measure $\mu_{k,\alpha}$ is related to $\mu_{1,\alpha}$ by Theorem 5. More precisely, we have the following corollary of Theorems 5 and 6:

Proposition. *Let* $f: \mathbb{Z}_p \to \mathbb{Z}_p$ *be the function* $f(x) = x^{k-1}$ *(k a fixed positive integer). Let X be a compact-open subset of $\mathbb{Z}_p$. Then*

$$\int_X 1\mu_{k,\alpha} = k \int_X f\mu_{1,\alpha}.$$

PROOF. By Theorem 5, we have

$$\mu_{k,\alpha}(a + (p^N)) \equiv ka^{k-1}\mu_{1,\alpha}(a + (p^N)) \pmod{p^{N-\operatorname{ord}_p d_k}}.$$

Now, assuming that N is large enough so that X is a union of intervals of the form $a + (p^N)$, we have

$$\int_X 1\mu_{k,\alpha} = \sum_{\substack{0 \le a < p^N \\ a+(p^N) \subset X}} \mu_{k,\alpha}(a + (p^N))$$

$$\equiv \sum_{\substack{0 \le a < p^N \\ a+(p^N) \subset X}} ka^{k-1}\mu_{1,\alpha}(a + (p^N)) \pmod{p^{N-\operatorname{ord}_p d_k}}$$

$$= k \sum_{\substack{0 \le a < p^N \\ a+(p^N) \subset X}} f(a)\mu_{1,\alpha}(a + (p^N)).$$

Taking the limit as $N \to \infty$ gives $\int_X 1\mu_{k,\alpha} = k\int_X f\mu_{1,\alpha}$ □

If we replace f by x^{k-1} in our notation, treating x as a "variable of integration," we may write this proposition as

$$\int_X 1\mu_{k,\alpha} = k \int_X x^{k-1}\mu_{1,\alpha}.$$

The right-hand side looks much better than the left hand side from the standpoint of *p*-adic interpolation, since instead of k appearing mysteriously in the subscript of μ, it appears in the exponent. And we know from §2 what the story is for interpolating the integrand x^{k-1} for any fixed x (see also Exercise 8 of §2). Namely, we're in business as long as $x \not\equiv 0 \pmod{p}$. To make all of our x's in the domain of integration have this property, we must take $X = \mathbb{Z}_p{}^\times$.

Thus, we claim that the expression $\int_{\mathbb{Z}_p{}^\times} x^{k-1}\mu_{1,\alpha}$ *can* be interpolated. To do this, we combine the results of §2 with the corollary at the end of §5. That corollary tells us that if $|f(x) - x^{k-1}|_p \leq \varepsilon$ for $x \in \mathbb{Z}_p{}^\times$, then

$$\left| \int_{\mathbb{Z}_p{}^\times} f\mu_{1,\alpha} - \int_{\mathbb{Z}_p{}^\times} x^{k-1}\mu_{1,\alpha} \right|_p \leq \varepsilon$$

(recall that $|\mu_{1,\alpha}(U)|_p \leq 1$ for all compact-open U). Choose for f the function $x^{k'-1}$, where $k' \equiv k \pmod{p-1}$ and $k' \equiv k \pmod{p^N}$ (writing this as one congruence: $k' \equiv k \pmod{(p-1)p^N}$). By §2, we have:

$$|x^{k'-1} - x^{k-1}|_p \leq \frac{1}{p^{N+1}} \quad \text{for } x \in \mathbb{Z}_p{}^\times.$$

Thus,

$$\left| \int_{\mathbb{Z}_p{}^\times} x^{k'-1}\mu_{1,\alpha} - \int_{\mathbb{Z}_p{}^\times} x^{k-1}\mu_{1,\alpha} \right|_p \leq \frac{1}{p^{N+1}}.$$

We conclude that, for any fixed $s_0 \in \{0, 1, 2, \ldots, p-2\}$, by letting k run through $S_{s_0} \underset{\text{def}}{=} \{\text{positive integers congruent to } s_0 \pmod{p-1}\}$, we can extend the function of k given by $\int_{\mathbb{Z}_p{}^\times} x^{k-1}\mu_{1,\alpha}$ to a continuous function of *p*-adic integers s:

$$\int_{\mathbb{Z}_p{}^\times} x^{s_0 + s(p-1) - 1}\mu_{1,\alpha}.$$

But we have strayed a little from our original numbers $-(1/k)\int_{\mathbb{Z}_p} 1\mu_{B,k}$. We just saw that we can interpolate

$$\int_{\mathbb{Z}_p{}^\times} x^{k-1}\mu_{1,\alpha} = \frac{1}{k}\int_{\mathbb{Z}_p{}^\times} 1\mu_{k,\alpha}.$$

Let's relate these two numbers:

$$\frac{1}{k}\int_{\mathbb{Z}_p{}^\times} 1\mu_{k,\alpha} = \frac{1}{k}\mu_{k,\alpha}(\mathbb{Z}_p{}^\times)$$

$$= \frac{1}{k}(1 - \alpha^{-k})(1 - p^{k-1})B_k \quad \text{(see Exercise 5 of §5)}$$

$$= (\alpha^{-k} - 1)(1 - p^{k-1})\left(-\frac{1}{k}\int_{\mathbb{Z}_p} 1\mu_{B,k}\right).$$

The term $1 - p^{k-1}$ made its appearance because we had to restrict our integration from $\mathbb{Z}_p$ to $\mathbb{Z}_p{}^\times$. This is the phenomenon predicted at the end of §2: because we can not interpolate n^s when $p|n$, we must remove a "*p*-Euler factor" from the ζ-function before it can be interpolated. So we will interpolate the numbers $(1 - p^{k-1})(-B_k/k)$:

$$(1 - p^{k-1})\left(-\frac{B_k}{k}\right) = \frac{1}{\alpha^{-k} - 1} \int_{\mathbb{Z}_p^\times} x^{k-1} \mu_{1,\alpha}.$$

One slight embarrassment, which we warned of in §2, is that the Euler term is $1 - p^{k-1}$ and not $1 - p^{-k}$ as you might think it should be from the heuristic discussion in §2. It's as though, instead of $\zeta(k)$, we were really interpolating "$\zeta(1 - k)$" (we haven't yet defined what this means for positive k). So we define our p-adic ζ-function to have the value $(1 - p^{k-1}) \cdot (-B_k/k)$ at the integer $1 - k$, not at k itself.

Definition. If k is a positive integer, let

$$\zeta_p(1 - k) \underset{\text{def}}{=} (1 - p^{k-1})(-B_k/k),$$

so that, by the preceding paragraph,

$$\zeta_p(1 - k) = \frac{1}{\alpha^{-k} - 1} \int_{\mathbb{Z}_p^\times} x^{k-1} \mu_{1,\alpha}.$$

Note that the expression on the right does not depend on α, i.e., if $\beta \in \mathbb{Z}$, $p \nmid \beta, \beta \neq 1$, then $(\beta^{-k} - 1)^{-1} \int_{\mathbb{Z}_p^\times} x^{k-1} \mu_{1,\beta} = (\alpha^{-k} - 1)^{-1} \int_{\mathbb{Z}_p^\times} x^{k-1} \mu_{1,\alpha}$, since both equal $(1 - p^{k-1})(-B_k/k)$. This equality—this independence of α—can also be proved directly (see Exercise 1). We shall use this independence of α later, when we define $\zeta_p(s)$ for p-adic s.

But we first derive some classical number theoretic facts about Bernoulli numbers. These facts were considered to be elegant but mysterious oddities until their connection with the Kubota–Leopoldt ζ_p and Mazur's measure $\mu_{1,\alpha}$ revealed them as natural outcomes of basic "calculus-type" considerations (namely, the corollary at the end of §5, which says, roughly speaking, that when two functions are close together on an interval, so are their integrals).

Theorem 7. (Kummer for (1) and (2), Clausen and von Staudt for (3)).

(1) If $p - 1 \nmid k$, then $|B_k/k|_p \leq 1$.

(2) If $p - 1 \nmid k$ and if $k \equiv k' \pmod{(p - 1)p^N}$, then

$$(1 - p^{k-1})\frac{B_k}{k} \equiv (1 - p^{k'-1})\frac{B_{k'}}{k'} \pmod{p^{N+1}}.$$

(3) If $p - 1 \mid k$ (or if $p = 2$ and k is even or $k = 1$), then

$$pB_k \equiv -1 \pmod{p}.$$

PROOF. We assume $p > 2$, and leave the proof of (3) when $p = 2$ as an exercise (Exercise 6 below).

We need a fact which will be proved at the beginning of the next chapter (at the end of §III.1): There exists an $\alpha \in \{2, 3, \ldots, p - 1\}$ such that α^{p-1} is the lowest positive power of α which is congruent to 1 modulo p. Put another way, the multiplicative group of nonzero residue classes of $\mathbb{Z}$ mopulo p is cyclic of order $p - 1$, i.e., there's a generator $\alpha \in \{2, 3, \ldots, p - 1\}$ such that the least positive residues of $\alpha, \alpha^2, \alpha^3, \ldots, \alpha^{p-1}$ exhaust $\{1, 2, 3, \ldots, p - 1\}$.

In the proof of parts (1) and (2), we choose our "measure regularizer" α to be such a generator in $\{2, 3, \ldots, p - 1\}$. This means that, since $p - 1 \nmid (-k)$, we have $\alpha^{-k} \not\equiv 1 \pmod{p}$, so that $(\alpha^{-k} - 1)^{-1} \in \mathbb{Z}_p^\times$.

To prove (1), we write (assuming $k > 1$; if $k = 1$ and $p > 2$, then $|B_1/1|_p = |-1/2|_p = 1$):

$$|B_k/k|_p = |1/(\alpha^{-k} - 1)|_p |1/(1 - p^{k-1})|_p \left| \int_{\mathbb{Z}_p^\times} x^{k-1}\mu_{1,\alpha} \right|_p$$

$$= \left| \int_{\mathbb{Z}_p^\times} x^{k-1}\mu_{1,\alpha} \right|_p$$

$$\leq 1,$$

by the proposition at the end of §5 (with $A = B = 1$), since $|\mu_{1,\alpha}(U)|_p \leq 1$ for all compact-open sets $U \subset \mathbb{Z}_p^\times$ and $|x^{k-1}|_p \leq 1$ for all $x \in \mathbb{Z}_p^\times$.

To prove (2), we rewrite the desired congruence as

$$\frac{1}{\alpha^{-k} - 1} \int_{\mathbb{Z}_p^\times} x^{k-1}\mu_{1,\alpha} \equiv \frac{1}{\alpha^{-k'} - 1} \int_{\mathbb{Z}_p^\times} x^{k'-1}\mu_{1,\alpha} \pmod{p^{N+1}}.$$

Notice that, if for $a, b, c, d \in \mathbb{Z}_p$ we have $a \equiv c \pmod{p^n}$ and $b \equiv d \pmod{p^n}$, then we also have $ab \equiv cb \equiv cd \pmod{p^n}$. Thus, since $a = (\alpha^{-k} - 1)^{-1}$, $b = \int_{\mathbb{Z}_p^\times} x^{k-1}\mu_{1,\alpha}$, $c = (\alpha^{-k'} - 1)^{-1}$, and $d = \int_{\mathbb{Z}_p^\times} x^{k'-1}\mu_{1,\alpha}$ are in $\mathbb{Z}_p$, it suffices to prove that $(\alpha^{-k} - 1)^{-1} \equiv (\alpha^{-k'} - 1)^{-1} \pmod{p^{N+1}}$ and $\int_{\mathbb{Z}_p^\times} x^{k-1}\mu_{1,\alpha} \equiv \int_{\mathbb{Z}_p^\times} x^{k'-1}\mu_{1,\alpha} \pmod{p^{N+1}}$. The first reduces to $\alpha^k \equiv \alpha^{k'} \pmod{p^{N+1}}$, and the second reduces (using the corollary at the end of §5, with $B = 1$ and $\varepsilon = p^{-N-1}$) to showing that $x^{k-1} \equiv x^{k'-1} \pmod{p^{N+1}}$ for all $x \in \mathbb{Z}_p^\times$. But this all follows from the discussion in §2.

Finally, we prove the Clausen-von Staudt congruence. For this let $\alpha = 1 + p$. Recall that we are proving it for $p > 2$. We have

$$pB_k = -kp(-B_k/k) = \frac{-kp}{\alpha^{-k} - 1}(1 - p^{k-1})^{-1} \int_{\mathbb{Z}_p^\times} x^{k-1}\mu_{1,\alpha}.$$

First take the first of the three terms on the right. If we let $d = \mathrm{ord}_p k$, then

$$\alpha^{-k} - 1 = (1 + p)^{-k} - 1 \equiv -kp \pmod{p^{d+2}},$$

so that

$$1 \equiv \frac{-kp}{\alpha^{-k} - 1} \pmod{p}.$$

Next, since k must be ≥ 2, we have $(1 - p^{k-1})^{-1} \equiv 1 \pmod{p}$. Thus,

$$pB_k \equiv \int_{\mathbb{Z}_p^\times} x^{k-1}\mu_{1,\alpha} \pmod{p}.$$

Again using the corollary at the end of §5, this time with $f(x) = x^{k-1}$ and $g(x) = 1/x$, we obtain

$$pB_k \equiv \int_{\mathbb{Z}_p^\times} x^{-1}\mu_{1,\alpha} \pmod{p}.$$

But by Exercise 7 of §5, this latter integral is congruent to $-1 \pmod{p}$. □

We now return to *p*-adic interpolation.

Definition. Fix $s_0 \in \{0, 1, 2, \ldots, p - 2\}$. For $s \in \mathbb{Z}_p$ ($s \neq 0$ if $s_0 = 0$), we define

$$\zeta_{p,s_0}(s) \underset{\text{def}}{\equiv} \frac{1}{\alpha^{-(s_0+(p-1)s)} - 1} \int_{\mathbb{Z}_p^\times} x^{s_0+(p-1)s-1}\mu_{1,\alpha}.$$

It should by now be clear that this definition makes sense, namely,

$$\alpha^{-(s_0+(p-1)s)} = \alpha^{-s_0}(\alpha^{p-1})^{-s} \quad \text{and} \quad x^{s_0+(p-1)s-1} \quad \text{for any } x \in \mathbb{Z}_p^\times$$

are defined for *p*-adic s by taking any sequence $\{k_i\}$ of positive integers which approach s *p*-adically. Another way to define $\zeta_{p,s_0}(s)$ is as follows: $-\lim_{k_i \to s}(1 - p^{s_0+(p-1)k_i-1})B_{s_0+(p-1)k_i}/(s_0 + (p - 1)k_i)$.

We now see that if k is a positive integer congruent to $s_0 \pmod{p - 1}$, i.e., $k = s_0 + (p - 1)k_0$, then we have: $\zeta_p(1 - k) = \zeta_{p,s_0}(k_0)$. We think of the ζ_{p,s_0} as *p*-adic "branches" of ζ_p, one for each congruence class mod $p - 1$. (But note that the odd congruence classes—$s_0 = 1, 3, \ldots, p - 2$—give us the zero function, since for such s_0 always $B_{s_0+(p-1)k_i} = 0$; so we are only interested in even s_0.)

In the definition of ζ_{p,s_0}, we excluded the case $s = 0$ when $s_0 = 0$. This is because in that case $\alpha^{-(s_0+(p-1)s)} = 1$, and the denominator vanishes. If we write $\zeta_p(1 - k) = \zeta_{p,s_0}(k_0)$, where $k = s_0 + (p - 1)k_0$, then this excluded case corresponds to $\zeta_p(1)$. Thus, the *p*-adic zeta-function, like the Archimedean Riemann zeta-function, has a "pole" at 1.

Theorem 8. *For fixed p and fixed s_0, $\zeta_{p,s_0}(s)$ is a continuous function of s which does not depend on the choice of $\alpha \in \mathbb{Z}$, $p{\nmid}\alpha$, $\alpha \neq 1$, which appears in its definition.*

PROOF. It is clear that §2 and the corollary at the end of §5 imply that the integral is a continuous function of s. The factor $1/(\alpha^{-(s_0+(p-1)s)} - 1)$ is a continuous function as long as we don't allow $s = 0$ when $s_0 = 0$, because $\alpha^{-(s_0+(p-1)s)}$ is a continuous function by §2. So $\zeta_{p,s_0}(s)$ is also continuous.

It remains to show that $\zeta_{p,s_0}(s)$ does not depend on α. Let $\beta \in \mathbb{Z}$, $p{\nmid}\beta$, $\beta \neq 1$. The two functions

$$\frac{1}{\alpha^{-(s_0+(p-1)s)} - 1} \int_{\mathbb{Z}_p^\times} x^{s_0+(p-1)s-1}\mu_{1,\alpha}$$

and

$$\frac{1}{\beta^{-(s_0+(p-1)s)} - 1} \int_{\mathbb{Z}_p^\times} x^{s_0+(p-1)s-1}\mu_{1,\beta}$$

agree whenever $s_0 + (p - 1)s = k$ is an integer greater than 0, i.e., whenever s is a nonnegative integer ($s > 0$ if $s_0 = 0$), since in that case both functions equal $(1 - p^{k-1})(-B_k/k)$. But the nonnegative integers are dense in $\mathbb{Z}_p$, so that any two continuous functions which agree there are equal. Therefore, taking β instead of α does not affect the function. $\square$

Theorem 8 gives us our p-adic interpolation of the "interesting factor" $-B_{2k}/2k$ in $\zeta(2k)$. But a few things remain to be explained: (1) the terminology "Mazur–Mellin transform" in the title of this section; and (2) the mysterious switch from k to $1 - k$. In addition, something should be said about (3) deeper analogues with classical ζ-functions and L-functions, and (4) a connection with modular forms. Since these four topics will take us beyond the scope of what we intend to prove in this book, they are gathered together in a section which surveys some basic relevant facts without attempting any proofs. References for proofs and further discussion of (1)–(4) are: (1) Manin, "Periods of cusp forms, and p-adic Hecke series," §8; (2) Iwasawa, *Lectures on p-adic L-functions*, §1 and appendix; (3) Iwasawa, *Lectures on p-adic L-functions*, especially §5, and Borevich and Shafarevich, *Number Theory*, p. 332–336; (4) Serre, "Formes modulaires et fonctions zêta p-adiques," in Springer Lectures Notes in Mathematics 350.

7. A brief survey (no proofs)

(1) For $s > 1$, $\zeta(s)$ can be expressed as an integral

$$\frac{1}{\Gamma(s)} \int_0^\infty x^{s-1} \frac{dx}{e^x - 1}.$$

where $\Gamma(s)$ is the gamma-function, which satisfies $\Gamma(s + 1) = s\Gamma(s)$, $\Gamma(1) = 1$, so that, in particular, $\Gamma(k) = (k - 1)!$ for positive integers k. (See Exercise 4 below for the case $s = k$.) The integral is what is known as a Mellin transform. For a function $f(x)$ defined on the positive reals, the function

$$g(s) = \int_0^\infty x^{s-1} f(x)\, dx,$$

whenever it exists, is called the Mellin transform of $f(x)$ (or of $f(x)\, dx$). Thus, $\Gamma(s)\zeta(s)$ is the Mellin transform of $dx/(e^x - 1)$, which exists for $s > 1$ (see Exercise 4 below).

In §6, we showed that the function which p-adically interpolates $(1 - p^{k-1})$ $(-B_k/k)$ is essentially (except for the $1/(\alpha^{-s} - 1)$ factor and the s_0 business)

$$\int_{\mathbb{Z}_p^\times} x^{s-1} \mu_{1,\alpha},$$

where $\mu_{1,\alpha}$ is the regularized Mazur measure. Thus, the p-adic ζ-function can be thought of, in analogy to the classical case, as the "*p*-adic Mazur–Mellin transform" of the regularized Mazur measure $\mu_{1,\alpha}$.

(2) If we consider

$$\zeta(s) = \sum_{n=1}^{\infty} \frac{1}{n^s}$$

for s complex, with real part > 1, this sum still converges and defines a "complex analytic" function of s. By the technique of "analytic continuation," $\zeta(s)$ can be extended onto the entire complex plane except for the point $s = 1$ (where its behavior is like the function $1/(s - 1)$). A very basic property of $\zeta(s)$ is that it satisfies a "functional equation" which relates its value at s to its value at $1 - s$. Namely,

$$\zeta(1 - s) = \frac{2 \cos(\pi s/2)\Gamma(s)}{(2\pi)^s} \zeta(s)$$

Let's let $s = 2k$ be a positive even integer. Then

$$\zeta(1 - 2k) = \frac{2 \cos \pi k \,(2k - 1)!}{(2\pi)^{2k}} \zeta(2k)$$

$$= \frac{2(-1)^k(2k - 1)!}{(2\pi)^{2k}} \frac{(-1)^k \pi^{2k} 2^{2k-1}}{(2k - 1)!} \left(-\frac{B_{2k}}{2k}\right) \quad \text{by Theorem 4}$$

$$= -\frac{B_{2k}}{2k}.$$

On the other hand, if s is an *odd* integer greater than 1, the right-hand side of the functional equation vanishes because $\cos(\pi s/2) = 0$ (we need $s > 1$ in order for $\zeta(s)$ to be finite). Hence $\zeta(1 - s)$ vanishes, and so there too $\zeta(1 - k) = -B_k/k$, but all this says is that $0 = 0$.

Table of $\zeta(1 - k) = -B_k/k$

$1 - k$	$\zeta(1 - k)$
-1	$-1/12$
-3	$1/120$
-5	$-1/252$
-7	$1/240$
-9	$-1/132$
-11	$691/32760$
-13	$-1/12$
-15	$3617/8160$
-17	$-43867/14364$
-19	$174611/6600$
-21	$-77683/276$

So what we were "really" interpolating was the Riemann ζ-function at *negative odd integers*. We can now summarize the relationship between ζ_p and ζ in the following simple way, using the definition of ζ_p:

$$\zeta_p(1 - k) = (1 - p^{k-1})\zeta(1 - k) \quad \text{for } k = 2, 3, 4, \ldots .$$

If we're a little sloppy (forgetting that everything will diverge), we can write: $\zeta(1 - k) = \prod_{\text{primes } q} 1/(1 - q^{k-1})$,

$$\zeta^*(1 - k) = \prod_{\text{primes } q, \, q \neq p} 1/(1 - q^{k-1}) = (1 - p^{k-1})\zeta(1 - k),$$

so the appearance of the $(1 - p^{k-1})$ factor makes heuristic sense from this devil-may-care point of view.

In the same tack, we can derive the formula $\zeta(1 - k) = -B_k/k$ in a completely straightforward way:

$$\zeta(1 - k) \underset{\text{def}}{=} \sum_{n=1}^{\infty} \frac{1}{n^{1-k}} = \sum_{n=1}^{\infty} n^{k-1}.$$

Since $(d/dt)^{k-1} e^{nt}|_{t=0} = n^{k-1}$, we may write

$$\zeta(1 - k) = \sum_{n=1}^{\infty} \left(\frac{d}{dt}\right)^{k-1} e^{nt} \bigg|_{t=0}$$

$$= \left(\frac{d}{dt}\right)^{k-1} \sum_{n=1}^{\infty} e^{nt} \bigg|_{t=0}$$

$$= \left(\frac{d}{dt}\right)^{k-1} \left(\frac{1}{1 - e^t} - 1\right) \bigg|_{t=0}$$

$$= \left(\frac{d}{dt}\right)^{k-1} \left(\frac{1}{1 - e^t}\right) \bigg|_{t=0}$$

$$= \left(\frac{d}{dt}\right)^{k-1} \left(-\frac{1}{t} \sum_{n=1}^{\infty} B_n \frac{t^n}{n!}\right) \bigg|_{t=0}$$

$$= \left(\frac{d}{dt}\right)^{k-1} \sum \left(-\frac{B_n}{n}\right) \frac{t^{n-1}}{(n-1)!} \bigg|_{t=0}$$

$$= -\frac{B_k}{k}.$$

(3) Connections between ζ_p and ζ go deeper. An important example requires us to consider the generalization of ζ to functions of the form

$$L_\chi(s) \underset{\text{def}}{=} \sum_{n=1}^{\infty} \frac{\chi(n)}{n^s}, \quad s > 0,$$

where χ is a "character" (see Exercises 9–10 of §2). As long as χ is not the "trivial" character (equal to 1 for all n), this function $L_\chi(s)$ converges when $s = 1$: $\sum_{n=1}^{\infty} (\chi(n)/n)$, and can be computed explicitly. The result is:

$$L_\chi(1) = -\frac{\tau(\chi)}{N} \sum_{a=1}^{N-1} \bar{\chi}(a) \log(1 - e^{-2\pi i a/N}),$$

where N is the conductor of χ and $\tau(\chi) = \sum_{a=1}^{N-1} \chi(a) e^{2\pi i a/N}$ (this formula easily reduces to those given in Exercises 9–10 of §2).

In a manner very similar to the construction of ζ_p, it is possible to interpolate $L_\chi(1 - k)$ by "p-adic L-functions" $L_{\chi,p}$. Amazingly, it turns out that $L_{\chi,p}(1)$ equals the following expression:

$$-\left(1 - \frac{\chi(p)}{p}\right) \frac{\tau(\chi)}{N} \sum_{a=1}^{N-1} \bar{\chi}(a) \log_p(1 - e^{-2\pi i a/N}),$$

in which "$\log_p$" is the "p-adic logarithm," which is a p-adic function of a p-adic variable (see §IV.1 and §IV.2), and all of the roots of unity that occur—namely, $e^{2\pi i a/N}$ and the values of χ—are considered as elements of an algebraic extension of $\mathbb{Q}_p$ (see §III.2–3). Here $(1 - (\chi(p)/p))$ should be thought of as the p-Euler factor (for the ζ-function, $\chi = 1$, and the Euler factor in $\zeta(1)$, if $\zeta(1)$ were finite, would be $(1 - (1/p))$; see also Exercise 9 of §2 concerning Euler products for L_χ). The rest of the expression for $L_{\chi,p}(1)$ is the same as $L_\chi(1)$, except that the classical log is replaced by its p-adic analogue $\log_p$.

(4) Very important in the study of elliptic curves and of modular forms (see Chapter VII of Serre, *A Course in Arithmetic*) are the Eisenstein series E_{2k}, $k \geq 2$, which are functions defined on all complex numbers z with positive imaginary part by:

$$E_{2k}(z) = -\frac{1}{2} \frac{B_{2k}}{2k} + \sum_{n=1}^{\infty} \sigma_{2k-1}(n) e^{2\pi i n z},$$

where

$$\sigma_m(n) \underset{\text{def}}{=} \sum_{d \mid n, d > 0} d^m.$$

The series should be thought of as a "Fourier series"—i.e., a series in powers of $e^{2\pi i z}$—with constant term equal to $\frac{1}{2}\zeta(1 - 2k)$.

It turns out that Eisenstein series can be p-adically interpolated. One hint of this is that we can interpolate each nth coefficient as long as $p \nmid n$. Namely, that coefficient $\sigma_{2k-1}(n)$ is a finite sum of functions d^{2k-1}, all of which can be interpolated, by §2, since $p \nmid d$. Then interpolating $\zeta(1 - 2k)$ can be thought of as "getting the constant coefficient too." Vague as this all may seem, it is actually possible to derive the results in this chapter using the theory of p-adic modular forms. For details, see Serre's paper mentioned before, and papers by N. Katz on p-adic Eisenstein measures and p-adic interpolation of Eisenstein series (see Bibliography).

EXERCISES

1. Using the relationship between $\mu_{1,\alpha}$, $\mu_{k,\alpha}$, and $\mu_{B,k}$, give a direct proof (without mentioning Bernoulli numbers) that

$$\frac{1}{\alpha^{-k} - 1} \int_{\mathbf{Z}_p} x^{k-1} \mu_{1,\alpha}$$

does not depend on α.

2. Check the Kummer congruences from the table of $\zeta(1 - k)$ when $p = 5$, $k = 2$, $k' = 22$, $N = 1$. Check in the table that the congruences fail when $p - 1 | k$. Use the Kummer congruences and the first few values of B_k to compute the following through the p^2-place:

(i) B_{102} in $\mathbf{Q}_5$ (ii) B_{296} in $\mathbf{Q}_7$ (iii) B_{592} in $\mathbf{Q}_7$.

3. Use Theorem 7 and Exercise 20 of §I.2 to prove the following version of the theorem of Clausen and von Staudt: $(B_k + \sum \frac{1}{p}) \in \mathbf{Z}$, for any even k (or $k = 1$), where the summation is taken over all p for which $p - 1 | k$.

4. Show that $\int_0^\infty x^{s-1} dx/(e^x - 1)$ exists when $s > 1$. By writing $1/(e^x - 1) = e^{-x}/(1 - e^{-x}) = \sum_{n=1}^\infty e^{-nx}$, prove that:

$$\int_0^\infty \frac{x^{k-1}}{e^x - 1} dx = (k - 1)! \, \zeta(k) \quad \text{for } k = 2, 3, 4, \ldots$$

(justify your computations).

5. Prove that

$$\int_0^\infty \frac{x^{k-1}}{e^x + 1} dx = (k - 1)! \, (1 - 2^{1-k}) \zeta(k) \quad \text{for } k = 2, 3, 4, \ldots .$$

Show that the function

$$\frac{1}{\Gamma(s)(1 - 2^{1-s})} \int_0^\infty \frac{x^{s-1}}{e^x + 1} dx,$$

which you just showed equals $\zeta(s)$ for $s = k = 2, 3, 4, \ldots$, exists and is continuous as a function of s for $s > 0$, $s \neq 1$.

6. Prove the Clausen-von Staudt theorem when $p = 2$.

CHAPTER III

Building up Ω

1. Finite fields

In what follows, we'll have to assume familiarity with a few basic notions concerning algebraic extensions of fields. It would take us too far afield to review all the proofs; for a complete and readable treatment, see Lang's *Algebra* or Herstein's *Topics in Algebra*. We shall need the following concepts and facts:

(1) The abstract definition of a *field F*; a field *extension K* of *F* is any field *K* containing *F*; a field extension *K* is called *algebraic* if every $\alpha \in K$ is an *algebraic element*, i.e., satisfies a polynomial equation with coefficients in *F*: $a_0 + a_1\alpha + a_2\alpha^2 + \cdots + a_n\alpha^n = 0$, where $a_i \in F$. For example, the set of all numbers $a + b\sqrt{2}$ with $a, b \in \mathbb{Q}$ is an algebraic extension of $\mathbb{Q}$.

(2) If *F* is any field, its *characteristic* char(*F*) is defined as the least *n* such that when you add 1 to itself *n* times you get 0. If $1 + 1 + \cdots + 1$ always $\neq 0$, we say char(*F*) = 0. (It might sound more sensible to say char(*F*) = ∞, but the convention is to say that such fields have characteristic 0.) $\mathbb{Q}$, $\mathbb{Q}_p$, $\mathbb{R}$, and $\mathbb{C}$ are fields of characteristic 0, while the set of residue classes modulo a prime *p* is a field of characteristic *p*. (We'll see more examples of fields of characteristic *p* in a little while.)

(3) The definition of a *vector space V* over a field *F*; what it means to have a *basis* for *V* over *F*; what it means for *V* to be *finite-dimensional*; if *V* is finite-dimensional, its *dimension* is the number of elements in a basis.

(4) A field extension *K* of *F* is an *F*-vector space; if it is finite-dimensional, it must be an algebraic extension, and its dimension is called the *degree* [*K*:*F*]. If $\alpha \in K$ has the property that every element of *K* can be written as a rational expression in α, we write $K = F(\alpha)$ and say that *K* is the extension obtained by "adjoining" α to *F*. If *K'* is a finite extension of *K*, then it is easy to see that *K'* is a finite extension of *F*, and $[K':F] = [K':K] \cdot [K:F]$.

(5) Any element α in a field extension K of F which is algebraic over F satisfies a *unique* monic irreducible polynomial over F ("monic" means it has leading coefficient 1, "irreducible" means it cannot be factored into a product of polynomials of lower degree with coefficients in F):

$$\alpha^n + a_{n-1}\alpha^{n-1} + \cdots + a_1\alpha + a_0 = 0, \qquad a_i \in F;$$

n is called the degree of α. The field extension $F(\alpha)$ has degree n over F (in fact, $\{1, \alpha, \alpha^2, \ldots, \alpha^{n-1}\}$ is one possible basis for $F(\alpha)$ as a vector space over F).

(6) If F is a field of characteristic 0 (for example, $\mathbb{Q}$ or $\mathbb{Q}_p$) or a finite field (we'll study finite fields in detail very soon), then it can be proved that any finite extension K of F is of the form $K = F(\alpha)$ for some $\alpha \in K$. α is called a "primitive element." (Actually, this holds if F is any "perfect" field, where "perfect" means that either char $(F) = 0$, or else, if char $(F) = p$, every element in F has a pth root in F.) Knowing a primitive element α of a field extension K makes it easier to study K, since it means that everything in K is a polynomial in α of degree $< n$, i.e., $K = \{\sum_{i=0}^{n-1} a_i\alpha^i \mid a_i \in F\}$.

(7) Given an irreducible polynomial f of degree n with coefficients in F, we can construct a field extension $K \supset F$ of degree n in which f has a root $\alpha \in K$. Roots of all possible polynomials with coefficients in F can be successively adjoined in this way to obtain an "algebraic closure" (written $F^{\text{alg cl}}$ or $\bar{F}$) of the field F; by definition, this means a smallest possible algebraically closed field containing F (recall: a field K is called algebraically closed if every polynomial with coefficients in K has a root in K). Any algebraic extension of F is contained in an algebraic closure of F (i.e., can be extended to an algebraic closure of F). Any two algebraic closures of F are isomorphic, so we usually say "the algebraic closure," meaning "any algebraic closure." The algebraic closure of a field F is usually the union of an infinite number of finite algebraic extensions of F; for example, the algebraic closure of $\mathbb{Q}$ consists of all complex numbers which satisfy a polynomial equation with rational coefficients. However, the algebraic closure of the real numbers $\mathbb{R}$ is $\mathbb{C} = \mathbb{R}(\sqrt{-1})$, which is a finite algebraic extension of $\mathbb{R}$ of degree 2; but this is the exception rather than the rule.

(8) If $K = F(\alpha)$, if K' is another extension field of F, and if $\sigma: K \to K'$ gives an isomorphism of K with a subfield of K' (where σ is an "F-homomorphism," i.e., it preserves the field operations, and $\sigma(a) = a$ for all $a \in F$), then the image $\sigma(\alpha)$ of α in K' satisfies the same monic irreducible polynomial over F as α does. Conversely, if $K = F(\alpha)$, if K' is another extension field of F, and if $\alpha' \in K'$ satisfies the same monic irreducible polynomial over F as α does, then there exists a unique isomorphism σ of K with the subfield $F(\alpha')$ of K' such that $\sigma(a) = a$ for all $a \in F$ and such that $\sigma(\alpha) = \alpha'$.

(9) In $\bar{F} = F^{\text{alg cl}}$, all the roots of the monic irreducible equation over F

satisfied by an element $\alpha \in \bar{F}$ are called the *conjugates* of α. There is a one-to-one correspondence between isomorphisms of $F(\alpha)$ with a subfield of $\bar{F}$ and conjugates α' of α (see the preceding paragraph (8)). If char $(F) = 0$ or if F is finite (or if F is perfect), then an irreducible polynomial can not have multiple roots, i.e., all the conjugates of α are distinct. In that case: (number of conjugates) $= [F(\alpha):F]$.

(10) If $K = F(\alpha)$, then K is called "Galois" if all of the conjugates of α are in K. In that case all conjugates of any $x = \sum_{i=0}^{n-1} a_i \alpha^i \in K$ are in K, since such a conjugate is of the form $\sum_{i=0}^{n-1} a_i \alpha'^i$, where α' is a conjugate of α. Examples of Galois extensions of $\mathbb{Q}$ are: $\mathbb{Q}(\sqrt{2})$ (since $\alpha = \sqrt{2}$ has one conjugate $\alpha' = -\sqrt{2}$, which is the other root of $x^2 - 2 = 0$; here $-\sqrt{2} \in \mathbb{Q}(\sqrt{2})$); $\mathbb{Q}(i)$; $\mathbb{Q}(\sqrt{d})$ for any $d \in \mathbb{Q}$; $\mathbb{Q}(\zeta_m)$, where $\zeta_m = e^{2\pi i/m}$ is a primitive mth root of 1 in $\mathbb{C}$ (since the conjugates of ζ_m are other primitive mth roots, and these are of the form ζ_m^i for i having no common factor with m). An example of a non-Galois extension of $\mathbb{Q}$ is $\mathbb{Q}(\sqrt[4]{2})$, since the conjugates of $\sqrt[4]{2}$ are the 4 roots of $x^4 - 2 = 0$, namely $\pm\sqrt[4]{2}$ and $\pm i\sqrt[4]{2}$, and we have $i\sqrt[4]{2} \notin \mathbb{Q}(\sqrt[4]{2})$ (since $\mathbb{Q}(\sqrt[4]{2})$ is contained in the real numbers).

(11) If K is a Galois extension of F, then the isomorphisms in paragraph (8) all have image K itself, i.e., they are F-isomorphisms from K to K, or "F-automorphisms of K." The set of these automorphisms is a group, called the "Galois group of K over F." If σ is such an automorphism, then the set of $x \in K$ such that $\sigma(x) = x$ is called the "fixed field of σ" (it's easy to see that it's a subfield of K containing F). For example, if $K = \mathbb{Q}(\sqrt{2} + \sqrt{3})$, which is a Galois field extension of $\mathbb{Q}$ of degree 4, and if σ takes $\sqrt{2} + \sqrt{3}$ to $\sqrt{2} - \sqrt{3}$, then the fixed field of σ turns out to be $\mathbb{Q}(\sqrt{2})$. It is not hard to prove that, if K is a Galois extension of F and $K' \neq K$ is a field between K and F: $F \subset K' \subset K$, then there is a nontrivial automorphism of K which leaves K' fixed. In turns out there is a one-to-one correspondence between subgroups S of the Galois group of K over F and such intermediate fields $F \subset K' \subset K$, where

$$S \leftrightarrow K_S' = \{x \in K \mid \sigma x = x \text{ for all } \sigma \in S\}.$$

But we shall only need simple cases of the facts in this paragraph, not the full power of Galois theory.

We now proceed to the study of finite fields. The simplest example of a finite field is the "integers modulo a prime p." This means: take the set of equivalence classes of integers for the equivalence relation: $x \sim y$ means $x \equiv y \pmod{p}$. There are p such equivalence classes: the class of 0, 1, 2, 3, $\ldots, p - 2, p - 1$. It is easy to define addition and multiplication and check that this set, which we call $\mathbb{F}_p$, forms a field (in particular, every non-zero equivalence class has an inverse; this amounts to saying that if p does not divide x, then there exists a y such that $xy \equiv 1 \pmod{p}$). $\mathbb{F}_p$ is sometimes written $\mathbb{Z}/p\mathbb{Z}$ (meaning "the integers divided out by p times the integers").

We could have equally well started out with the p-adic integers $\mathbb{Z}_p$, and defined $x \sim y$ $(x, y \in \mathbb{Z}_p)$ to mean $x \equiv y \pmod{p}$ (i.e., x and y have the same first digit in their p-adic expansions). That is, $\mathbb{F}_p$ can also be written $\mathbb{Z}_p/p\mathbb{Z}_p$ ("the p-adic integers divided out by p times the p-adic integers"). $\mathbb{Z}_p/p\mathbb{Z}_p$ is called the "residue field" of $\mathbb{Z}_p$. The reason why we have to study general finite fields before going further is as follows: the residue fields we get when, instead of $\mathbb{Q}_p$ and $\mathbb{Z}_p$, we deal with algebraic extensions of $\mathbb{Q}_p$, are not quite as simple as $\mathbb{F}_p$. They turn out to be algebraic extensions of $\mathbb{F}_p$. So we need to get a picture of what general finite fields look like.

Let F be a finite field. Since not all the numbers $0, 1, 1 + 1, 1 + 1 + 1, \ldots$ can be distinct, F must have characteristic $\neq 0$. Let $n = \operatorname{char}(F)$. Note that n must be a prime, since if we could write $n = n_0 n_1$, with n_0 and n_1 both $< n$, we would have $n_0 \neq 0$, so multiplying by n_0^{-1} would give: $n_1 = n_0^{-1} n = 0$, a contradiction. So let p denote the prime number $\operatorname{char}(F)$.

Clearly, any field F of characteristic p contains the field of p elements as a subfield (namely, by taking the subfield of F formed by all numbers of the form $1 + \cdots + 1$). This subfield is called the "prime field" of F.

Note that in any field F of characteristic p, the map $x \mapsto x^p$ preserves addition and multiplication:

$$xy \mapsto (xy)^p = x^p y^p;$$

$$x + y \mapsto (x + y)^p = \sum_{i=0}^{p} \binom{p}{i} x^i y^{p-i} = x^p + y^p,$$

because for $1 \le i \le p - 1$ the integer $\binom{p}{i} = p!/(i! \, (p - i)!)$ is divisible by p, and hence equal to 0 in F.

Theorem 9. *Let F be a finite field containing q elements, and let $f = [F:\mathbb{F}_p]$ (i.e., the dimension of F as a vector space over its prime field $\mathbb{F}_p$). Let K be an algebraic closure of $\mathbb{F}_p$ containing F. Then $q = p^f$; F is the only field of q elements contained in K; and F is the set of all elements of K satisfying the equation $x^q - x = 0$. Conversely, for any power $q = p^f$ of p, the roots of the equation $x^q - x = 0$ in K are a field of q elements.*

PROOF. Since F is an f-dimensional vector space over $\mathbb{F}_p$, the number of elements is equal to the number of choices of the f components (i.e., "coordinates" in terms of a basis of f elements) from $\mathbb{F}_p$, which is p^f. Next, any field F of q elements has $q - 1$ nonzero elements, so that the nonzero elements of F under multiplication form a group of order $q - 1$. In this group the powers of an element x form a subgroup of order equal to the least power of x which equals one (called the "order" of x). But it is easy to prove that any subgroup of a finite group has order dividing the order of the group. Thus, x has order dividing $q - 1$, and so $x^{q-1} = 1$ for all nonzero x in F. Then $x^q - x = 0$ for all x (including 0) in F. Since this holds for *any* field of q elements in K, and a polynomial of degree q has at most q distinct roots in a field, it follows that any field of q elements in K must be the roots of $x^q - x$, and there is only one such set of q roots.

Conversely, given any $q = p^f$, the set of elements of K such that $x^q = x$ is closed under addition and multiplication (the argument is the same as in the paragraph right before the statement of this theorem), and so is a subfield of K. This polynomial has distinct roots, because, if it had a double root, by Exercise 10 below, that root would also be a root of the formal derivative polynomial $qx^{q-1} - 1 = -1$ (because $q = 0$ in K); but the polynomial -1 has no roots. □

Remark. Because any two algebraic closures of $\mathbb{F}_p$ are isomorphic, it follows that any two fields of $q = p^f$ elements are isomorphic.

We let $\mathbb{F}_q$ denote the unique (up to isomorphism) field of $q = p^f$ elements.

If F is a field, $F^\times$ denotes the multiplicative group of non-zero elements of F.

Proposition. $\mathbb{F}_q^\times$ *is a cyclic group of order* $q - 1$.

PROOF. If we let $o(x)$ denote the order of x (the least power of x which equals 1), we know that $o(x)$ is a divisor of $q - 1$ for all $x \in \mathbb{F}_q^\times$. But if d is any divisor of $q - 1$, the equation $x^d = 1$ has at most d solutions, because the degree d polynomial $x^d - 1$ has at most d roots in a field. If $d = o(x)$, then all d distinct elements $x, x^2, \ldots, x^{d-1}, x^d = 1$ satisfy this equation, and so they must be the only ones that do. How many of these d elements have order *exactly* d? It is easy to see that the answer is: the number of integers in $\{1, 2, \ldots, d - 1, d\}$ which are relatively prime to d (have no common divisor with d other than 1). This number is denoted $\varphi(d)$. Thus, at most $\varphi(d)$ elements of $\mathbb{F}_q^\times$ have order d. We claim that exactly $\varphi(d)$ have order d for all divisors d of $q - 1$, and in particular for $d = q - 1$. This follows from the following lemma.

Lemma. $\sum_{d|n} \varphi(d) = n$.

PROOF OF THE LEMMA. Let $\mathbb{Z}/n\mathbb{Z}$ denote the additive group $\{0, 1, \ldots, n - 1\}$ of integers modulo n. $\mathbb{Z}/n\mathbb{Z}$ contains a subgroup S_d for each divisor d of n defined as follows: S_d is the set of all multiples of n/d. Clearly, every subgroup of $\mathbb{Z}/n\mathbb{Z}$ is obtained in this way.

S_d has d elements, of which $\varphi(d)$ generate the full subgroup (i.e., the set of all multiples of mn/d exhausts the set of all multiples of n/d in $\mathbb{Z}/n\mathbb{Z}$ if and only if m and d are relatively prime). But each integer $0, 1, \ldots, n - 1$ generates one of the subgroups S_d. Hence

$$\{0, 1, \ldots, n - 1\} = \bigcup_{d|n} \{\text{elements generating } S_d\}.$$

Since this is a disjoint union, we have: $n = \sum_{d|n} \varphi(d)$, and the lemma is proved.

The proposition follows immediately, because if there were fewer than $\varphi(d)$ elements of order d for some $d|n$, we would have: $n = \sum_{d|n} \{\text{elements of order } d\} < \sum_{d|n} \varphi(d) = n$. Hence, in particular, there are $\varphi(q - 1)$ elements of order $q - 1$. Since $\varphi(q - 1) \geq 1$ (for example, 1 has no common factor with $q - 1$), there exists an element a of order exactly $q - 1$. Then $\mathbb{F}_q{}^\times = \{a, a^2, \ldots, a^{q-1}\}$. □

EXERCISES

1. Let F be a field of $q = p^f$ elements. Show that F contains one (and only one) field of $q' = p^{f'}$ elements if and only if f' divides f.

2. For $p = 2, 3, 5, 7, 11$, and 13, find an element $a \in \{1, 2, \ldots, p - 1\}$ which generates $\mathbb{F}_p{}^\times$, i.e., such that $\mathbb{F}_p{}^\times = \{a, a^2, \ldots, a^{p-1}\}$. In each case, determine how many choices there are for such an element a.

3. Let F be the set of numbers of the form $a + bj$, where $a, b \in \mathbb{F}_3 = \{0, 1, 2\}$, addition is defined component-wise, and multiplication is defined by $(a + bj)(c + dj) = (ac + 2bd) + (ad + bc)j$. Show that $F = \mathbb{F}_9$; show that $1 + j$ is a generator of $\mathbb{F}_9{}^\times$; and find all possible choices of a generator of $\mathbb{F}_9{}^\times$.

4. Write $\mathbb{F}_4$ and $\mathbb{F}_8$ explicitly in the same way as was done for $\mathbb{F}_9$ in the previous problem. Explain why any element except 1 in $\mathbb{F}_4{}^\times$ or $\mathbb{F}_8{}^\times$ is a generator.

5. Let $q = p^f$, and let a be an element generating $\mathbb{F}_q{}^\times$. Let $P(X)$ be the monic irreducible polynomial which a satisfies over $\mathbb{F}_p$. Prove that $\deg P = f$.

6. Let $q = p^f$. Prove that there are precisely f automorphisms of $\mathbb{F}_q$ over $\mathbb{F}_p$, namely the automorphisms $\sigma_i, i = 0, 1, \ldots, f - 1$, given by: $\sigma_i(x) = x^{p^i}$ for $x \in \mathbb{F}_q$.

7. Let $\alpha \in \mathbb{F}_p{}^\times$, and let $P(X) = X^p - X - \alpha$. Show that, if a is a root of $P(X)$, then so is $a + 1, a + 2$, etc. Show that the field obtained by adjoining a to $\mathbb{F}_p$ has degree p over $\mathbb{F}_p$, i.e., it is isomorphic to $\mathbb{F}_{p^p}$.

8. Prove that $\mathbb{F}_q$ contains a square root of -1 if and only if $q \not\equiv 3 \pmod 4$.

9. Let ξ be algebraic of degree n over $\mathbb{Q}_p$, i.e., ξ satisfies a polynomial equation of degree n with coefficients in $\mathbb{Q}_p$, but none of degree less than n. Prove that there exists an integer N such that ξ does not satisfy any congruence

$$a_{n-1}\xi^{n-1} + a_{n-2}\xi^{n-2} + \cdots + a_1\xi + a_0 \equiv 0 \pmod{p^N},$$

in which the a_i are rational integers not all of which are divisible by p.

10. If F is *any* field, and $f(X) = X^n + a_{n-1}X^{n-1} + \cdots + a_1 X + a_0$ has co-efficients in F and factors in F, i.e., $f(X) = \prod_{i=1}^{n} (X - \alpha_i)$ with $\alpha_i \in F$, show that any root α_i which occurs more than once is also a root of $nX^{n-1} + a_{n-1}(n - 1)X^{n-2} + a_{n-2}(n-2)X^{n-3} + \cdots + a_1$.

2. Extension of norms

If X is a metric space, we say X is *compact* if every sequence has a convergent subsequence (see beginning of §II.3). For example, $\mathbb{Z}_p$ is a compact metric space (see Exercise 20 of §I.5). We say that X is *locally compact* if every point

$x \in X$ has a neighborhood (i.e., a subset of X containing some disc $\{y \mid d(x, y) < \epsilon\}$) which is compact. The real numbers $\mathbb{R}$ are not compact, but are a locally compact metric space with the usual Archimedian absolute value metric. $\mathbb{Q}_p$ with the p-adic metric is another example of a locally compact metric space, for the simple reason that for any x the neighborhood $x + \mathbb{Z}_p \underset{\text{def}}{=} \{y \mid |y - x|_p \leq 1\}$ is compact (in fact, it is isomorphic to $\mathbb{Z}_p$ as a metric space). More generally, if X is an additive group such that $d(x, y) = d(x - y, 0)$ for all x, y (for example, if X is a vector space and the metric is induced from a norm on X, as defined below), then X is locally compact whenever 0 has a compact neighborhood U. Namely, for any x, the translation of U by x: $x + U \underset{\text{def}}{=} \{y \mid y - x \in U\}$, is a compact neighborhood of x. In $\mathbb{Q}_p$, $U = \mathbb{Z}_p$ is such a compact neighborhood of 0. It is not hard to see (Exercise 6 below) that any such locally compact group is complete.

Let F be a field with a non-Archimedean norm $\| \ \|$. For the duration of this section we assume that F is locally compact.

Let V be a finite dimensional vector space over F. By a norm on V we mean the analogous thing to a norm on a field, namely, a map $\| \ \|_V$ from V to the nonnegative real numbers satisfying: (1) $\|x\|_V = 0$ if and only if $x = 0$; (2) $\|ax\|_V = \|a\| \, \|x\|_V$ for all $x \in V$ and $a \in F$ (here $\|a\|$ is the norm in F); and (3) $\|x + y\|_V \leq \|x\|_V + \|y\|_V$. For example, if K is a finite extension field of F, then any norm on K as a field whose restriction to F is $\| \ \|$ is also a norm on K as a vector space. However, a word of caution: the converse is false, since Property (2) for a vector space norm is weaker than the corresponding property for a field norm (see Exercises 3–4 below).

As in the case of fields, we say that two norms $\| \ \|_1$ and $\| \ \|_2$ on V are equivalent if a sequence of vectors is Cauchy with respect to $\| \ \|_1$ if and only if it is Cauchy with respect to $\| \ \|_2$. This is true if and only if there exist positive constants c_1 and c_2 such that for all $x \in V$: $\|x\|_2 \leq c_1 \|x\|_1$ and $\|x\|_1 \leq c_2 \|x\|_2$ (see Exercise 1 below).

Theorem 10. *If V is a finite dimensional vector space over a locally compact field F, then all norms on V are equivalent.*

PROOF. Let $\{v_1, \ldots, v_n\}$ be a basis for V. Define the sup-norm $\| \ \|_{\sup}$ (pronounced "soup norm") on V by

$$\|a_1 v_1 + \cdots + a_n v_n\|_{\sup} \underset{\text{def}}{=} \max_{1 \leq i \leq n} (\|a_i\|).$$

Note that this definition depends on the choice of basis. This $\| \ \|_{\sup}$ is a norm (see Exercise 2 below). Now let $\| \ \|_V$ be any other norm on V. First of all, for any $x = a_1 v_1 + \cdots + a_n v_n$ we have

$$\|x\|_V \leq \|a_1\| \, \|v_1\|_V + \cdots + \|a_n\| \, \|v_n\|_V$$
$$\leq n \, (\max \|a_i\|) \max \|v_i\|_V,$$

so that we get $\| \ \|_V \leq c_1 \| \ \|_{\sup}$ if we choose $c_1 = n \max_{1 \leq i \leq n}(\|v_i\|_V)$. It remains to find a constant c_2 such that the reverse inequality holds; then it will follow that any norm on V is equivalent to the sup-norm.

Let

$$U = \{x \in V \mid \|x\|_{\text{sup}} = 1\}.$$

We claim that there is some positive ε such that $\|x\|_V \geq \varepsilon$ for $x \in U$. If this weren't the case, we would have a sequence $\{x_i\}$ in U such that $\|x_i\|_V \to 0$. By the compactness of U with respect to $\| \ \|_{\text{sup}}$ (Exercises 2, 8 below), there exists a subsequence $\{x_{i_j}\}$ which converges in the sup-norm to some $x \in U$. But for every j

$$\|x\|_V \leq \|x - x_{i_j}\|_V + \|x_{i_j}\|_V \leq c_1 \|x - x_{i_j}\|_{\text{sup}} + \|x_{i_j}\|_V,$$

by our first inequality relating the two norms. Both of these terms approach 0 as $j \to \infty$, since x_{i_j} converges to x in $\| \ \|_{\text{sup}}$, and $\|x_i\|_V \to 0$. Hence $\|x\|_V = 0$, so that $x = 0 \notin U$, a contradiction.

Using this claim, we can easily prove the second inequality, and hence the theorem. The idea is: the claim says that on the sup-norm unit sphere U the other norm $\| \ \|_V$ remains greater than or equal to some positive number ε, and hence $\| \ \|_{\text{sup}} \leq c_2 \| \ \|_V$ on U, where $c_2 = 1/\varepsilon$ (on U the left side of this inequality is 1, by definition); but everything in V can be obtained by multiplying U by scalars (elements in F), so the same inequality holds on all of V.

More precisely, let $x = a_1 v_1 + \cdots + a_n v_n$ be any nonzero element in V, and choose j so that $\|a_j\| = \max \|a_i\| = \|x\|_{\text{sup}}$. Then clearly $(x/a_j) \in U$, and so

$$\|x/a_j\|_V \geq \varepsilon = 1/c_2,$$

so that

$$\|x\|_{\text{sup}} = \|a_j\| \leq c_2 \|x\|_V. \qquad \square$$

Corollary. *Let $V = K$ be a field. Then there is at most one norm $\| \ \|_K$ of K as a field which extends $\| \ \|$ on F (i.e., such that $\|a\|_K = \|a\|$ for $a \in F$).*

PROOF OF COROLLARY. Note that the field norm $\| \ \|_K$ is also an F-vector space norm, because it extends $\| \ \|$. By Theorem 10, any two such norms $\| \ \|_1$ and $\| \ \|_2$ on K must be equivalent. Hence $\| \ \|_2 \leq c_1 \| \ \|_1$. Let $x \in K$ be such that $\|x\|_1 \neq \|x\|_2$, say, $\|x\|_1 < \|x\|_2$. But then for a sufficiently large N we have $c_1 \|x^N\|_1 < \|x^N\|_2$, a contradiction. $\qquad \square$

This still leaves the question of whether there exists *any* norm on K extending $\| \ \|$ on F.

We now recall a basic concept in field extensions, that of the "norm" of an element. This use of the word "norm" should not be confused with the use so far in the sense of metrics. "Norm" in the new sense will always be in quotation marks and denoted by $\mathbb{N}$.

Let $K = F(\alpha)$ be a finite extension of a field F generated by an element α which satisfies a monic irreducible equation

$$0 = x^n + a_1 x^{n-1} + \cdots + a_{n-1} x + a_n, \qquad a_i \in F.$$

Then the following three definitions of the "norm of α from K to F," abbreviated $\mathsf{N}_{K/F}(\alpha)$, are equivalent:

(1) If K is considered as an n-dimensional vector space over F, then multiplication by α is an F-linear map from K to K having some matrix A_α. We let $\mathsf{N}_{K/F}(\alpha) \underset{\text{def}}{=} \det(A_\alpha)$.

(2) $\mathsf{N}_{K/F}(\alpha) \underset{\text{def}}{=} (-1)^n a_n$.

(3) $\mathsf{N}_{K/F}(\alpha) \underset{\text{def}}{=} \prod_{i=1}^n \alpha_i$, where the α_i are the conjugates of $\alpha = \alpha_1$ over F.

The equivalence (2) $\Leftrightarrow$ (3) comes from: $x^n + a_1 x^{n-1} + \cdots + a_n = \prod_{i=1}^n (x - \alpha_i)$. The equivalence (1) $\Leftrightarrow$ (2) is easy to see if we use $\{1, \alpha, \alpha^2, \ldots, \alpha^{n-1}\}$ as a basis for K over F. Namely, the matrix of multiplication by α is then clearly (using: $\alpha^n = -a_1 \alpha^{n-1} - \cdots - a_{n-1}\alpha - a_n$):

$$\begin{pmatrix} 0 & 0 & & & & -a_n \\ 1 & 0 & 0 & & & -a_{n-1} \\ & 1 & 0 & & & \\ & & & \ddots & & \vdots \\ & & & & 0 & -a_2 \\ & & & & 1 & -a_1 \end{pmatrix}$$

which has determinant $(-1)^n a_n$, as follows immediately by expanding using the first row.

If $\beta \in K = F(\alpha)$, we can define $\mathsf{N}_{K/F}(\beta)$ as either (1) the determinant of the matrix of multiplication by β in K, or, equivalently, (2) $(\mathsf{N}_{F(\beta)/F}(\beta))^{[K:F(\beta)]}$. The two are equivalent because, if we choose bases for $F(\beta)$ as a vector space over F and for K as a vector space over $F(\beta)$, then as a basis for K over F we can take all products of an element in the first basis with an element in the second basis; using this basis for K over F, we see that the matrix of multiplication by β in K takes the following "block form"

$$\begin{pmatrix} A_\beta & 0 & & \\ 0 & A_\beta & & \\ & & \ddots & \\ & & & A_\beta \end{pmatrix},$$

where A_β is the matrix of multiplication by β in $F(\beta)$. The determinant of this matrix is the $[K:F(\beta)]$-th power ($[K:F(\beta)]$ is the number of blocks) of $\det A_\beta$, i.e., the $[K:F(\beta)]$-th power of $\mathsf{N}_{F(\beta)/F}(\beta)$. Thus, the two definitions of $\mathsf{N}_{K/F}(\beta)$ are in fact equivalent.

Since $\mathsf{N}_{K/F}(\alpha)$ is defined for *any* $\alpha \in K$ as the determinant of the matrix of multiplication by α in K, it follows that $\mathsf{N}_{K/F}$ is a *multiplicative* map from K to F, i.e., $\mathsf{N}_{K/F}(\alpha\beta) = \mathsf{N}_{K/F}(\alpha)\mathsf{N}_{K/F}(\beta)$. (Namely, multiplication by $\alpha\beta$ is given by the product of the matrix for α and the matrix for β, and the determinant of a product of matrices is the product of the determinants.)

We can now figure out how the extension of $|\ |_p$ to an algebraic number $\alpha \in \mathbb{Q}_p^{\text{alg cl}}$ must be defined if it exists. Suppose α has degree n, i.e., its monic irreducible polynomial over $\mathbb{Q}_p$ has degree n. Let K be a finite *Galois* extension of $\mathbf{Q}_p$ containing α (see paragraph (10) at the beginning of this chapter), for example, K can be the field obtained by adjoining α and all of its conjugates to $\mathbf{Q}_p$ (it's easy to check that this field is finite and Galois over $\mathbf{Q}_p$). Suppose we find an extension $\| \ \|$ of $| \ |_p$ to K. By the corollary to Theorem 10, $\| \ \|$ is the *unique* field norm on K extending $| \ |_p$. Now let α' be any conjugate of α, and let σ be a $\mathbf{Q}_p$-automorphism of K taking α to α' (see paragraphs (8), (9), and (11) in §III.1). Clearly the map $\| \ \|' : K \to \mathbb{R}$ defined by $\|x\|' = \|\sigma(x)\|$ is a field norm on K which extends $| \ |_p$. Hence $\| \ \|' = \| \ \|$, and so $\|\alpha\| = \|\alpha\|' = \|\sigma(\alpha)\| = \|\alpha'\|$. We conclude that *the norm of α equals the norm of each of its conjugates*. But then the norm of $N_{\mathbf{Q}_p(\alpha)/\mathbf{Q}_p}(\alpha)$, which is in $\mathbf{Q}_p$, equals

$$
\begin{aligned}
|N_{\mathbf{Q}_p(\alpha)/\mathbf{Q}_p}(\alpha)|_p &= \|N_{\mathbf{Q}_p(\alpha)/\mathbf{Q}_p}(\alpha)\| \\
&= \| \prod_{\text{conjugates } \alpha' \text{ of } \alpha} \alpha' \| \\
&= \prod \|\alpha'\| \\
&= \|\alpha\|^n.
\end{aligned}
$$

Thus,

$$
\|\alpha\| = |N_{\mathbf{Q}_p(\alpha)/\mathbf{Q}_p}(\alpha)|_p^{1/n}.
$$

So, concretely speaking, to find the p-adic norm of α, look at the monic irreducible polynomial satisfied by α over $\mathbf{Q}_p$. If it has degree n and constant term a_n, then the p-adic norm of α is the nth root of $|a_n|_p$. (Of course, we have not yet proved that this rule really has all of the required properties of a norm; this will be Theorem 11 below.)

Note that we can equivalently define $\|\alpha\|$ to be

$$
|N_{K/\mathbf{Q}_p}(\alpha)|_p^{1/[K:\mathbf{Q}_p]},
$$

where K is now any field of finite degree over $\mathbf{Q}_p$ that contains α,

$$
N_{K/\mathbf{Q}_p}(\alpha) = (N_{\mathbf{Q}_p(\alpha)/\mathbf{Q}_p}(\alpha))^{[K:\mathbf{Q}_p(\alpha)]},
$$

and

$$
n = [\mathbf{Q}_p(\alpha):\mathbf{Q}_p] = \frac{[K:\mathbf{Q}_p]}{[K:\mathbf{Q}_p(\alpha)]}.
$$

We now prove that this rule $\| \ \|$ really is a norm. We shall write $| \ |_p$ instead of $\| \ \|$ to denote the extension of $| \ |_p$ to K; this should not cause confusion. The reader should be warned that Theorem 11 is not an easy fact to prove. The proof given here, which was told to me by D. Kazhdan, is much more efficient than other proofs I've seen. But it should be read and re-read carefully until the reader is thoroughly convinced by the argument.

Theorem 11. *Let K be a finite extension of $\mathbf{Q}_p$. Then there exists a field norm on K which extends the norm $| \ |_p$ on $\mathbf{Q}_p$.*

PROOF. Let $n = [K:\mathbf{Q}_p]$. We first define $| \ |_p$ on K, and then prove that it's really a field norm on K extending $| \ |_p$ on $\mathbf{Q}_p$. For any $\alpha \in K$ we define

$$
|\alpha|_p \underset{\text{def}}{=} |N_{K/\mathbf{Q}_p}(\alpha)|_p^{1/n},
$$

where the right-hand side is the old norm in $\mathbb{Q}_p$. (Here n is the degree of the field K over $\mathbb{Q}_p$, and is not necessarily the degree of the element α over $\mathbb{Q}_p$.) It is easy to check that: (1) $|\alpha|_p$ agrees with the old $|\alpha|_p$ whenever $\alpha \in \mathbb{Q}_p$; (2) $|\alpha|_p$ is multiplicative; and (3) $|\alpha|_p = 0 \Leftrightarrow \alpha = 0$. The hard part is the property: $|\alpha + \beta|_p \leq \max(|\alpha|_p, |\beta|_p)$.

Suppose that $|\beta|_p$ is the larger of $|\alpha|_p, |\beta|_p$. Setting $\gamma = \alpha/\beta$, we have $|\gamma|_p \leq 1$. We want to show that $|\alpha + \beta|_p \leq \max(|\alpha|_p, |\beta|_p) = |\beta|_p$, or equivalently (after dividing through by $|\beta|_p$): $|1 + \gamma|_p \leq 1$. Thus, Theorem 11 follows from the following lemma.

Lemma. *With* $| \ |_p$ *defined as above on* K, *one has* $|1 + \gamma|_p \leq 1$ *for any* $\gamma \in K$ *with* $|\gamma|_p \leq 1$.

PROOF OF THE LEMMA. We noted before that we can define $|\gamma|_p$ and $|1 + \gamma|_p$ using the field $\mathbb{Q}_p(\gamma) = \mathbb{Q}_p(1 + \gamma)$ in place of K:

$$|\gamma|_p = |N_{\mathbb{Q}_p(\gamma)/\mathbb{Q}_p}(\gamma)|_p^{1/[\mathbb{Q}_p(\gamma):\mathbb{Q}_p]}; \ |1 + \gamma|_p = |N_{\mathbb{Q}_p(\gamma)/\mathbb{Q}_p}(1 + \gamma)|_p^{1/[\mathbb{Q}_p(\gamma):\mathbb{Q}_p]}.$$

So without loss of generality we may suppose that $K = \mathbb{Q}_p(\gamma)$, in other words, that γ is a "primitive element" of K. Then $\{1, \gamma, \gamma^2, \ldots, \gamma^{n-1}\}$ is a vector space basis for K over $\mathbb{Q}_p$, where $n = [K:\mathbb{Q}_p]$.

For any element $\alpha = \sum_{i=0}^{n-1} a_i \gamma^i \in K$, let $\|\alpha\|$ denote the sup-norm in this basis, i.e., $\|\alpha\| \underset{\text{def}}{=} \max_i |a_i|_p$. Similarly, if $A = \{a_{ij}\}$ is any $n \times n$ matrix with entries in $\mathbb{Q}_p$, let $\|A\|$ denote the sup-norm $\|A\| \underset{\text{def}}{=} \max_{i,j} |a_{ij}|_p$.

Any $\mathbb{Q}_p$-linear map from K to K, when written in terms of the basis $\{1, \gamma, \gamma^2, \ldots, \gamma^{n-1}\}$, gives an $n \times n$ matrix with entries in $\mathbb{Q}_p$. Now let A denote the matrix of the $\mathbb{Q}_p$-linear map from K to K which is multiplication by γ. (This is the type of matrix used before in our discussion of the three equivalent definitions of the norm of an element.) Then the matrix A^i is the matrix corresponding to multiplication by γ^i, and $I + A$ is the matrix corresponding to multiplication by $1 + \gamma$. (More generally, the matrix $P(A)$ corresponds to multiplication by the element $P(\gamma)$ for any polynomial $P \in \mathbb{Q}_p[X]$.)

We claim that the sequence of real numbers $\{\|A^i\|\}_{i=0,1,2,\ldots}$ is bounded above. Suppose the contrary. Then we can find a sequence i_j, $j = 1, 2, \ldots$, such that $\|A^{i_j}\| > j$. Let $b_j \underset{\text{def}}{=} \|A^{i_j}\|$, which is the maximum $| \ |_p$ of any of the n^2 entries of A^{i_j}. Let β_j be an entry of A^{i_j} with maximum $| \ |_p$; thus, $|\beta_j|_p = \|A^{i_j}\| = b_j$. Define the matrix $B_j = A^{i_j}/\beta_j$, i.e., divide all entries of A^{i_j} by β_j. (Note that $\beta_j \neq 0$, since $\|A^{i_j}\| > j$.) Then clearly $\|B_j\| = 1$. Since the sup-norm unit ball is compact (Exercises 2 and 8 below), we can find a subsequence $\{B_{j_k}\}_{k=1,2,\ldots}$ which converges to some matrix B. Since $\det B_j = (\det A^{i_j})/\beta_j^n$, we have

$$|\det B_j|_p < |\det A^{i_j}|_p/j^n = |N_{K/\mathbb{Q}_p}(\gamma)^{i_j}|_p/j^n = |\gamma|_p^{n i_j}/j^n \leq 1/j^n.$$

By the definition of convergence in the sup-norm, each entry of B is the limit as $k \to \infty$ of the corresponding entry of B_{j_k}; hence $\det B = \lim \det B_{j_k} = 0$.

Because $\det B = 0$, there exists a nonzero element $l \in K$, considered as a vector written with respect to the basis $\{1, \gamma, \gamma^2, \ldots, \gamma^{n-1}\}$, such that $Bl = 0$. We now show that this implies that B is identically zero, contradicting $\|B\| = 1$ and hence proving the claim that $\{\|A^i\|\}$ is bounded.

Since $\{\gamma^i l\}_{i=0,1,\dots,n-1}$ is a basis for the $\mathbb{Q}_p$-vector space K, it suffices to show that $B\gamma^i l = 0$ for any i. But since multiplication by γ^i is given by the matrix A^i, we have $B\gamma^i l = (BA^i)l = A^i Bl = 0$, where the relation $BA^i = A^i B$ comes from the fact that B is a limit of matrices of the form $B_j = A^{i_j}/\beta_j$, i.e., scalar multiples of powers of A, and any such matrix B_j commutes with A^i. This proves our claim that $\{\|A^i\|\}$ is bounded by some constant C.

Note that for any $n \times n$ matrix $A = \{a_{ij}\}$ we have: $|\det A|_p \le (\max_{i,j}|a_{ij}|_p)^n = \|A\|^n$; this is clear if we expand the determinant and use the additive and multiplicative properties of a non-Archimedean norm.

Now let N be very large, and consider: $(I + A)^N = I + \binom{N}{1}A + \cdots + \binom{N}{N-1}A^{N-1} + A^N$. We have

$$|1 + \gamma|_p^N = |\det(I + A)^N|_p^{1/n} \le \|(I + A)^N\| \le \left(\max_{0 \le i \le N} \left\| \binom{N}{i}A^i \right\| \right)$$
$$\le \left(\max_{0 \le i \le N} \|A^i\| \right) \le C.$$

Hence $|1 + \gamma|_p \le \sqrt[N]{C}$. Letting $N \to \infty$ gives $|1 + \gamma|_p \le 1$ as required. (Note the similarity with the proof of Ostrowski's theorem in §I.2.) $\qquad\square$

Let R be a (commutative) *ring*, i.e., a set R with two operations $+$ and $\cdot$ which satisfy all the rules of a field except for the existence of multiplicative inverses. In other words, it's an additive group under $+$; has associativity, identity, and commutativity under $\cdot$; and has distributivity. R is called an *integral domain* if $xy = 0$ always implies $x = 0$ or $y = 0$. $\mathbb{Z}$ and $\mathbb{Z}_p$ are examples of integral domains.

A *proper* subset I of R is called an *ideal* if it is an additive subgroup of R and for all $x \in R$ and $a \in I$ we have: $xa \in I$. In the ring $\mathbb{Z}$, the set of all multiples of a fixed integer is an ideal. In $\mathbb{Z}_p$, for any $r \le 1$ the set $\{x \in \mathbb{Z}_p | \ |x|_p < r\}$ is an ideal. If, say, $r = p^{-n}$, this is the set of all p-adic integers whose first $n + 1$ digits are zero in the p-adic expansion.

If I_1 and I_2 are ideals of R, then the set

$$\{x \in R \mid x \text{ can be written as } x = x_1 x_1' + \cdots + x_m x_m' \text{ with } x_i \in I_1, x_i' \in I_2\}$$

is easily checked to be an ideal, which is written $I_1 I_2$ and is called the product of the two ideals. An ideal I is called *prime* if: $x_1 x_2 \in I$ implies $x_1 \in I$ or $x_2 \in I$.

It is easy to verify (see Exercise 5 below) that $\mathbb{Z}_p$ has precisely one prime ideal, namely

$$p\mathbb{Z}_p \underset{\text{def}}{=} \{x \in \mathbb{Z}_p \mid |x|_p < 1\},$$

and that all ideals of $\mathbb{Z}_p$ are of the form

$$p^n\mathbb{Z}_p \underset{\text{def}}{=} \{x \in \mathbb{Z}_p \mid |x|_p \le p^{-n}\}.$$

If I is an ideal in a ring R, it is easy to see that the set of additive cosets $x + I$ form a ring, called R/I. (Another way of describing this ring: the set of equivalence classes of elements of R with respect to the equivalence relation

$x \sim y$ if $x - y \in I$.) For example, if $R = \mathbb{Z}$ (or if $R = \mathbb{Z}_p$), the ring R/pR is the field $\mathbb{F}_p$ of p elements, as we've seen.

An ideal M in R is called maximal if there is no ideal strictly between M and R. It is an easy exercise to check that:

(1) An ideal P is prime if and only if R/P is an integral domain.
(2) An ideal M is maximal if and only if R/M is a field.

Now let K be a finite extension field of $\mathbb{Q}_p$. (Or, more generally, let K be an algebraic extension of a field F which is the field of fractions of an integral domain R, e.g., $F = \mathbb{Q}$ is the field of fractions of $R = \mathbb{Z}$, $F = \mathbb{Q}_p$ is the field of fractions of $R = \mathbb{Z}_p$.) Let A be the set of all $\alpha \in K$ which satisfy an equation of the form $x^n + a_1 x^{n-1} + \cdots + a_{n-1}x + a_n = 0$ with the $a_i \in \mathbb{Z}_p$. (Every $\alpha \in K$ of course satisfies an equation of this form with coefficients in $\mathbb{Q}_p$, but usually not all the a_i are in $\mathbb{Z}_p$.) A is called the "integral closure of $\mathbb{Z}_p$ in K."

It is not hard to show that if $\alpha \in A$, then its monic irreducible polynomial has the above form. Moreover, the integral closure is always a ring. (For the general proof, see Lang's *Algebra*, pp. 237–240; in the case we'll be working with—the integral closure of $\mathbb{Z}_p$ in K—we prove that it's a ring in the proposition that follows.)

Proposition. *Let K be a finite extension of $\mathbb{Q}_p$ of degree n, and let*

$$A = \{x \in K \mid |x|_p \leq 1\},$$
$$M = \{x \in K \mid |x|_p < 1\}.$$

Then A is a ring, which is the integral closure of $\mathbb{Z}_p$ in K. M is its unique maximal ideal, and A/M is a finite extension of $\mathbb{F}_p$ of degree at most n.

PROOF. It is easy to check that A is a ring and M is an ideal in A, using the additive and multiplicative properties of a non-Archimedean norm. Now let $\alpha \in K$ have degree m over $\mathbb{Q}_p$, and suppose that α is integral over $\mathbb{Z}_p$: $\alpha^m + a_1\alpha^{m-1} + \cdots + a_m = 0$, $a_i \in \mathbb{Z}_p$. If $|\alpha|_p > 1$, we would have:

$$|\alpha|_p^m = |\alpha^m|_p = |a_1\alpha^{m-1} + \cdots + a_m|_p \leq \max_{1 \leq i \leq m} |a_i\alpha^{m-i}|_p$$

$$\leq \max_{1 \leq i \leq m} |\alpha^{m-i}|_p = |\alpha|_p^{m-1},$$

a contradiction. Conversely, suppose $|\alpha|_p \leq 1$. Then all the conjugates of $\alpha = \alpha_1$ over $\mathbb{Q}_p$ also have $|\alpha_i|_p = \prod_{j=1}^m |\alpha_j|_p^{1/m} = |\alpha|_p \leq 1$. Since all the coefficients in the monic irreducible polynomial of α are sums or differences of products of α_i (the so-called "symmetric polynomials" in the α_i), it follows that these coefficients also have $|\ |_p \leq 1$. Since they lie in $\mathbb{Q}_p$, they hence must lie in $\mathbb{Z}_p$.

We now prove that M contains every ideal of A. Suppose $\alpha \in A$, $\alpha \notin M$. Then $|\alpha|_p = 1$, so that $|1/\alpha|_p = 1$, and $1/\alpha \in A$. Hence any ideal containing α must contain $(1/\alpha) \cdot \alpha = 1$, which is impossible.

Note that $M \cap \mathbb{Z}_p = p\mathbb{Z}_p$ from the definition of M.

Consider the field A/M. Recall that its elements are cosets $a + M$. Notice that if a and b happen to be in $\mathbb{Z}_p$, then $a + M$ is the same coset as $b + M$ if and only if $a - b \in M \cap \mathbb{Z}_p = p\mathbb{Z}_p$. Thus, there's a natural inclusion of $\mathbb{Z}_p/p\mathbb{Z}_p$ into A/M given by coset $a + p\mathbb{Z}_p \mapsto$ coset $a + M$ for $a \in \mathbb{Z}_p$. Since $\mathbb{Z}_p/p\mathbb{Z}_p$ is the field $\mathbb{F}_p$ of p elements, this means that A/M is an extension field of $\mathbb{F}_p$.

We now claim that A/M has finite degree over $\mathbb{F}_p$, in fact, that $[A/M : \mathbb{F}_p] \leq [K : \mathbb{Q}_p]$. If $n = [K : \mathbb{Q}_p]$, we show that any $n + 1$ elements $\bar{a}_1, \bar{a}_2, \ldots, \bar{a}_{n+1} \in A/M$ must be linearly dependent over $\mathbb{F}_p$. For $i = 1, 2, \ldots, n + 1$, let a_i be any element in A which maps to $\bar{a}_i$ under the map $A \to A/M$ (i.e., a_i is any element in the coset $\bar{a}_i$, in other words: $\bar{a}_i = a_i + M$). Since $[K : \mathbb{Q}_p] = n$, it follows that $a_1, a_2, \ldots, a_{n+1}$ are linearly dependent over $\mathbb{Q}_p$:

$$a_1 b_1 + a_2 b_2 + \cdots + a_{n+1} b_{n+1} = 0, \qquad b_i \in \mathbb{Q}_p.$$

Multiplying through by a suitable power of p, we may assume that all the $b_i \in \mathbb{Z}_p$ but at least one b_i is not in $p\mathbb{Z}_p$. Then the image of this expression in A/M is

$$\bar{a}_1 \bar{b}_1 + \bar{a}_2 \bar{b}_2 + \cdots + \bar{a}_{n+1} \bar{b}_{n+1} = 0,$$

where $\bar{b}_i$ is the image of b_i in $\mathbb{Z}_p/p\mathbb{Z}_p$ (i.e., $\bar{b}_i$ is the first digit in the p-adic expansion of b_i). Since at least one b_i is not in $p\mathbb{Z}_p$, it follows that at least one $\bar{b}_i$ is not 0, so that $\bar{a}_1, \bar{a}_2, \ldots, \bar{a}_{n+1}$ are linearly dependent, as claimed. $\square$

The field A/M is called the *residue field* of K. It's a field extension of $\mathbb{F}_p$ of some finite degree f. A itself is called the "valuation ring" of $| \ |_p$ in K.

EXERCISES

1. Prove that two vector space norms $\| \ \|_1$ and $\| \ \|_2$ on a finite dimensional vector space V are equivalent if and only if there exist $c_1 > 0$ and $c_2 > 0$ such that for all $x \in V$:
$$\|x\|_2 \leq c_1 \|x\|_1 \quad \text{and} \quad \|x\|_1 \leq c_2 \|x\|_2.$$

2. Let F be a field with a norm $\| \ \|$. Let V be a finite dimensional vector space over F with a basis $\{v_1, \ldots, v_n\}$. Prove that $\|a_1 v_1 + \cdots + a_n v_n\|_{\text{sup}} \underset{\text{def}}{=} \max_{1 \leq i \leq n} (\|a_i\|)$ is a norm on V. Prove that if F is locally compact, then V is locally compact with respect to $\| \ \|_{\text{sup}}$.

3. Let $V = \mathbb{Q}_p(\sqrt{p})$, $v_1 = 1$, $v_2 = \sqrt{p}$. Show that the sup-norm is not a *field* norm on $\mathbb{Q}_p(\sqrt{p})$.

4. If $V = K$ is a field, can the sup-norm *ever* be a field norm (for any basis $\{v_1, \ldots, v_n\}$) when $n = \dim K > 1$? Discuss what type of finite extensions K of $\mathbb{Q}_p$ can never have the sup-norm being a field norm.

5. Prove that $\mathbb{Z}_p$ has precisely one prime ideal, namely $p\mathbb{Z}_p$, and that all ideals in $\mathbb{Z}_p$ are of the form $p^n \mathbb{Z}_p$, $n \in \{1, 2, 3, \ldots\}$.

6. Prove that, if a vector space with a norm $\| \ \|_V$ is locally compact, then it is complete.

7. Prove that a vector space with a norm $\| \ \|_V$ is locally compact if and only if $\{x \mid \|x\|_V \leq 1\}$ is compact.

8. Prove that, if a vector space with a norm $\| \ \|_V$ is locally compact, then $\{x \mid \|x\|_V = 1\}$ is compact.

3. The algebraic closure of $\mathbb{Q}_p$

Putting together the two theorems in §2, we conclude that $| \ |_p$ has a unique extension (which we also denote $| \ |_p$) to any finite field extension of $\mathbb{Q}_p$. Since the algebraic closure $\bar{\mathbb{Q}}_p$ of $\mathbb{Q}_p$ is the union of such extensions, $| \ |_p$ extends uniquely to $\bar{\mathbb{Q}}_p$. Concretely speaking, if $\alpha \in \bar{\mathbb{Q}}_p$ has monic irreducible polynomial $x^n + a_1 x^{n-1} + \cdots + a_n$, then $|\alpha|_p = |a_n|_p^{1/n}$.

Let K be an extension of $\mathbb{Q}_p$ of degree n. For $\alpha \in K^\times$ we define

$$\text{ord}_p \, \alpha \underset{\text{def}}{=} -\log_p |\alpha|_p = -\log_p |N_{K/\mathbb{Q}_p}(\alpha)|_p^{1/n} = -\frac{1}{n} \log_p |N_{K/\mathbb{Q}_p}(\alpha)|_p.$$

This agrees with the earlier definition of ord_p when $\alpha \in \mathbb{Q}_p$, and clearly has the property that $\text{ord}_p \, \alpha\beta = \text{ord}_p \, \alpha + \text{ord}_p \, \beta$. Note that the definition of $\text{ord}_p \, \alpha$ is independent of the choice of field K such that $\alpha \in K$, $[K : \mathbb{Q}_p] < \infty$. The image of $K^\times$ under the ord_p map is contained in $(1/n)\mathbb{Z} \underset{\text{def}}{=} \{x \in \mathbb{Q} \mid nx \in \mathbb{Z}\}$. Since this image is a nontrivial additive subgroup of $(1/n)\mathbb{Z}$ that contains $\mathbb{Z}$, it must be of the form $(1/e)\mathbb{Z}$ for some positive integer e dividing n. This integer e is called the "index of ramification" of K over $\mathbb{Q}_p$. If $e = 1$, we say that K is an *unramified* extension of $\mathbb{Q}_p$. Now let $\pi \in K$ be any element such that $\text{ord}_p \, \pi = (1/e)$. Then clearly any $x \in K$ can be written uniquely in the form

$$\pi^m u, \quad \text{where } |u|_p = 1 \text{ and } m \in \mathbb{Z} \text{ (in fact, } m = e \cdot \text{ord}_p \, x).$$

It can be proved (Exercise 12 below) that $n = e \cdot f$, where $n = [K : \mathbb{Q}_p]$, e is the index of ramification, and f is the degree of the residue field A/M over $\mathbb{F}_p$. In any case, we've already seen that $f \leq n$ and $e \leq n$. In the case of an unramified extension K. i.e., when $e = 1$, we may choose p itself for π in the preceding paragraph, since $\text{ord}_p \, p = 1 = (1/e)$. At the other extreme, if $e = n$, the extension K is called *totally ramified*.

Proposition. *If K is totally ramified and $\pi \in K$ has the property $\text{ord}_p \, \pi = (1/e)$, then π satisfies an "Eisenstein equation" (see Exercise 14 of §I.5)*

$$x^e + a_{e-1} x^{e-1} + \cdots + a_0 = 0, \qquad a_i \in \mathbb{Z}_p,$$

where $a_i \equiv 0 \pmod p$ for all i, and $a_0 \not\equiv 0 \pmod{p^2}$. Conversely, if α is a root of such an Eisenstein equation over $\mathbb{Q}_p$, then $\mathbb{Q}_p(\alpha)$ is totally ramified over $\mathbb{Q}_p$ of degree e.

PROOF. Since the a_i are symmetric polynomials in the conjugates of π, all of which have $| \ |_p = p^{-1/e}$, it follows that $|a_i|_p < 1$. As for a_0, we have $|a_0|_p = |\pi|_p^e = 1/p$.

Conversely, we saw in Exercise 14 §I.5 that an Eisenstein polynomial is irreducible, so that adjoining a root α gives us an extension of degree e.

Since $\text{ord}_p \, a_0 = 1$, it follows that $\text{ord}_p \, \alpha = (1/e) \, \text{ord}_p \, a_0 = (1/e)$, and hence $\mathbb{Q}_p(\alpha)$ is totally ramified over $\mathbb{Q}_p$. $\qquad\square$

A more precise description of the types of roots of polynomials that can be used to get a totally ramified extension of degree e can be given if e is not divisible by p (this case is called "tame" ramification; $p|e$ is called "wild" ramification). Namely, such tamely totally ramified extensions are obtained by adjoining solutions of the equation $x^e - pu = 0$, where $u \in \mathbb{Z}_p{}^\times$, i.e., such extensions are always obtained by extracting an eth root of p times a p-adic unit (see Exercises 13 and 14 below).

Now let K be any finite extension of $\mathbb{Q}_p$. The next proposition tells us that if K is unramified, i.e., $e = 1$, then K must be of a very special type, namely, a field obtained by adjoining a root of 1; while if K is ramified, it can be obtained by first adjoining a suitable root of 1 to obtain its "maximal unramified subfield" and then adjoining to this subfield a root of an Eisenstein polynomial. *Warning:* the proof of the following proposition is slightly tedious, and the reader who is impatient to get to the meatier material in the next chapter may want to skip it (and also skip over some of the harder exercises in §III.4) on a first reading.

Proposition. *There is exactly one unramified extension K_f^{unram} of $\mathbb{Q}_p$ of degree f, and it can be obtained by adjoining a primitive $(p^f - 1)$th root of 1. If K is an extension of $\mathbb{Q}_p$ of degree n, index of ramification e, and residue field degree f (so that $n = ef$, as proved in Exercise 12 below), then $K = K_f^{\text{unram}}(\pi)$, where π satisfies an Eisenstein polynomial with coefficients in K_f^{unram}.*

PROOF. Let $\bar{a}$ be a generator of the multiplicative group $\mathbb{F}_{p^f}^\times$ (see the proposition at the end of §1), and let $\bar{P}(x) = x^f + \bar{a}_1 x^{f-1} + \cdots + \bar{a}_f$, $\bar{a}_i \in \mathbb{F}_p$, be its monic irreducible polynomial over $\mathbb{F}_p$ (see Exercise 5 of §1). For each i, let $a_i \in \mathbb{Z}_p$ be any element which reduces to $\bar{a}_i \bmod p$, and let $P(x) = x^f + a_1 x^{f-1} + \cdots + a_f$. Clearly, $P(x)$ is irreducible over $\mathbb{Q}_p$, since otherwise it could be written as a product of two polynomials with coefficients in $\mathbb{Z}_p$, and each could be reduced mod p to get $\bar{P}(x)$ as a product. Let $\alpha \in \mathbb{Q}_p^{\text{alg cl}}$ be a root of $P(x)$. Let $\tilde{K} = \mathbb{Q}_p(\alpha)$, $\tilde{A} = \{x \in K \mid |x|_p \leq 1\}$, $\tilde{M} = \{x \in K \mid |x|_p < 1\}$. Then $[\tilde{K} : \mathbb{Q}_p] = f$, while the coset $\alpha + \tilde{M}$ satisfies the degree f irreducible polynomial $\bar{P}(x)$ over $\mathbb{F}_p$. Hence $[\tilde{A}/\tilde{M} : \mathbb{F}_p] = f$, and $\tilde{K}$ is an unramified extension of degree f. (We have not yet proved that it is the only one.)

Now let $\tilde{K}$ be as in the second part of the proposition. Let $A = \{x \in \tilde{K} \mid |x|_p \leq 1\}$ be the valuation ring of $|\ |_p$ in $\tilde{K}$, and let $M = \{x \in \tilde{K} \mid |x|_p \leq 1\}$ be the maximal ideal of A, so that $A/M = \mathbb{F}_{p^f}$. Let $\bar{a} \in \mathbb{F}_{p^f}$ be a generator of the multiplicative group $\mathbb{F}_{p^f}^\times$. Let $\alpha_0 \in A$ be any element that reduces to $\bar{a} \bmod M$. Finally, let $\pi \in \tilde{K}$ be any element with $\text{ord}_p \, \pi = 1/e$; thus, $M = \pi A$.

We claim that there exists $\alpha \equiv \alpha_0 \bmod \pi$ such that $\alpha^{p^f - 1} - 1 = 0$. The proof is a Hensel's lemma type argument. Namely, we write $\alpha \equiv \alpha_0 + \alpha_1 \pi$

(mod π^2), so that mod π^2 we need $0 \equiv (\alpha_0 + \alpha_1\pi)^{p^f-1} - 1 \equiv \alpha_0^{p^f-1} - 1 + (p^f - 1)\alpha_1\pi\alpha_0^{p^f-2} \equiv \alpha_0^{p^f-1} - 1 - \alpha_1\pi\alpha_0^{p^f-2}$ (mod π^2). But $\alpha_0^{p^f-1} \equiv 1$ (mod π), so that, if we set $\alpha_1 \equiv (\alpha_0^{p^f-1} - 1)/(\pi\alpha_0^{p^f-2})$ (mod π), then we get the desired congruence mod π^2. Continuing in this way, just as in Hensel's lemma, we find a solution $\alpha = \alpha_0 + \alpha_1\pi + \alpha_2\pi^2 + \cdots$ to the equation $\alpha^{p^f-1} = 1$.

Note that $\alpha, \alpha^2, \ldots, \alpha^{p^f-1}$ are all distinct, because their reductions mod M—$\bar{\alpha}, \bar{\alpha}^2, \ldots, \bar{\alpha}^{p^f-1}$—are distinct. In other words, α is a primitive $(p^f - 1)$th root of 1. Also note that $[\mathbb{Q}_p(\alpha):\mathbb{Q}_p] \geq f$, since f is the residue field degree of the extension. (We will soon prove that $[\mathbb{Q}_p(\alpha):\mathbb{Q}_p] = f$.)

The above discussion applies, in particular, to the field $\tilde{K}$ constructed in the first paragraph of the proof. Hence $\tilde{K} \supset \mathbb{Q}_p(\alpha)$, where α is a primitive $(p^f - 1)$th root of 1. Since $f = [\tilde{K}:\mathbb{Q}_p] \geq [\mathbb{Q}_p(\alpha):\mathbb{Q}_p] \geq f$, it follows that $\tilde{K} = \mathbb{Q}_p(\alpha)$. Thus, the unramified extension of degree f is unique. Call it $\tilde{K}_f^{\mathrm{unram}}$.

We now return to our field K of degree $n = ef$ over $\mathbb{Q}_p$. Let $E(x)$ be the monic irreducible polynomial of π over $\tilde{K} = K_f^{\mathrm{unram}}$. Let $\{\pi_i\}$ be the conjugates of π over K_f^{unram}, so that $E(x) = \prod (x - \pi_i)$. Let d be the degree and c the constant term of $E(x)$. Then $\mathrm{ord}_p c = d\,\mathrm{ord}_p \pi = d/e$. But since $ef = n = [K:\mathbb{Q}_p] = [K:K_f^{\mathrm{unram}}][K_f^{\mathrm{unram}}:\mathbb{Q}_p] = [K:K_f^{\mathrm{unram}}]\cdot f$, it follows that $d \leq e$. Since $c \in K_f^{\mathrm{unram}}$, $\mathrm{ord}_p c$ is an integer. We conclude that $d = e$, and $\mathrm{ord}_p c = 1$. Thus, $E(x)$ is an Eisenstein polynomial, and $K = K_f^{\mathrm{unram}}(\pi)$. $\square$

Corollary. *If K is a finite extension of $\mathbb{Q}_p$ of degree n, index of ramification e, and residue field degree f, and if π chosen so that $\mathrm{ord}_p \pi = 1/e$, then every $\alpha \in K$ can be written in one and only one way as*

$$\sum_{i=m}^{\infty} a_i\pi^i,$$

where $m = e\,\mathrm{ord}_p \alpha$ and each a_i satisfies $a_i^{p^f} = a_i$ (i.e., the a_i's are Teichmüller digits).

The proof of the corollary is easy, and will be left to the reader.

If m is any positive integer not divisible by p, we can find a power p^f of p which is congruent to 1 mod m (namely, let f be the order of p in the multiplicative group $(\mathbb{Z}/m\mathbb{Z})^\times$ of residue classes mod m of integers prime to m). Then, if $p^f - 1 = mm'$, and if we adjoin to $\mathbb{Q}_p$ a primitive $(p^f - 1)$th root α of 1, it follows that $\alpha^{m'}$ is a primitive mth root of 1. Hence, we may conclude that *finite unramified extensions of $\mathbb{Q}_p$ are precisely the extensions obtained by adjoining roots of 1 of order not divisible by p.*

The union of all the finite unramified extensions of $\mathbb{Q}_p$ is written $\mathbb{Q}_p^{\mathrm{unram}}$ and is called the "maximal unramified extension of $\mathbb{Q}_p$." The ring of integers $\mathbb{Z}_p^{\mathrm{unram}}$ of $\mathbb{Q}_p^{\mathrm{unram}}$ (also called the "valuation ring"), which is

$$\mathbb{Z}_p^{\mathrm{unram}} \underset{\mathrm{def}}{=} \{x \in \mathbb{Q}_p^{\mathrm{unram}} \mid |x|_p \leq 1\},$$

has a (unique) maximal ideal $M^{\mathrm{unram}} = p\mathbb{Z}_p^{\mathrm{unram}} = \{x \in \mathbb{Q}_p^{\mathrm{unram}} \mid |x|_p < 1\} = \{x \in \mathbb{Q}_p^{\mathrm{unram}} \mid |x|_p \le 1/p\}$. The residue field $\mathbb{Z}_p^{\mathrm{unram}}/p\mathbb{Z}_p^{\mathrm{uuram}}$ is easily seen to be the algebraic closure $\bar{\mathbb{F}}_p$ of $\mathbb{F}_p$. Every $\bar{x} \in \bar{\mathbb{F}}_p$ has a unique "Teichmüller representative" $x \in \mathbb{Z}_p^{\mathrm{unram}}$ which is a root of 1 and has image $\bar{x}$ in $\mathbb{Z}_p^{\mathrm{unram}}/p\mathbb{Z}_p^{\mathrm{unram}}$. For this reason, $\mathbb{Z}_p^{\mathrm{unram}}$ is often called the "lifting to characteristic zero of $\bar{\mathbb{F}}_p$," (also called the "Witt vectors of $\bar{\mathbb{F}}_p$").

$\mathbb{Q}_p^{\mathrm{unram}}$, which is a much smaller field than $\mathbb{Q}_p^{\mathrm{alg\,cl}}$, can be used instead of $\mathbb{Q}_p^{\mathrm{alg\,cl}}$ in many situations.

The "opposite" of unramified extensions is totally ramified extensions. We can get a totally ramified extension, for example, by adjoining a primitive p^rth root of 1—this will give us a totally ramified extension of degree $n = e = p^{r-1}(p - 1)$ (see Exercise 7 below). However, unlike in the unramified case, not by a long shot can all totally ramified extensions be obtained by adjoining roots of 1. For example, adjoining a root of $x^m - p$ clearly gives a totally ramified extension K of degree m; but if K were contained in the field obtained by adjoining a primitive p^rth root of 1, we would have $m \mid p^{r-1}(p - 1)$, which is impossible if, say, $m > p$ and $p \nmid m$. About all we can say about the set of all totally ramified extensions of $\mathbb{Q}_p$ is contained in the proposition at the beginning of this section and in Exercise 14 below.

We repeat: An extension K of $\mathbb{Q}_p$ of degree n, index of ramification e, and residue field degree f is obtained by adjoining a primitive $(p^f - 1)$th root of 1 and then adjoining to the resulting field K_f^{unram} a root of an Eisenstein polynomial with coefficients in K_f^{unram}.

We conclude this section with a couple of useful propositions.

Proposition (Krasner's Lemma). *Let $a, b \in \bar{\mathbb{Q}}_p$ $(= \mathbb{Q}_p^{\mathrm{alg\,cl}})$, and assume that b is chosen closer to a than all conjugates a_i of a $(a_i \ne a)$, i.e.,*

$$|b - a|_p < |a_i - a|_p.$$

Then $\mathbb{Q}_p(a) \subset \mathbb{Q}_p(b)$.

PROOF. Let $K = \mathbb{Q}_p(b)$, and suppose $a \notin K$. Then, since a has conjugates *over K* equal in number to $[K(a):K]$, which is > 1, it follows that there is at least one $a_i \notin K$, $a_i \ne a$, and there is an isomorphism σ of $K(a)$ to $K(a_i)$ which keeps K fixed and takes a to a_i. We already know, because of the uniqueness of the extension of norms, that $|\sigma x|_p = |x|_p$ for all $x \in K(a)$. In particular, $|b - a_i|_p = |\sigma b - \sigma a|_p = |b - a|_p$, and hence

$$|a_i - a|_p \le \max(|a_i - b|_p, |b - a|_p) = |b - a|_p < |a_i - a|_p,$$

a contradiction. $\qquad\square$

Note that Krasner's Lemma can be proved in exactly the same way in a more general situation: If $a, b \in \bar{\mathbb{Q}}_p$, K is a finite extension of $\mathbb{Q}_p$, and for all conjugates a_i of a over K ($a_i \neq a$) we have $|b - a|_p < |a_i - a|_p$, then $K(a) \subset K(b)$.

Now let K be any field with a norm $\| \ \|$. If f, $g \in K[X]$, i.e., $f = \sum a_i X^i$ and $g = \sum b_i X^i$ are two polynomials with coefficients in K, we define the distance $\|f - g\|$ from f to g as

$$\|f - g\| \underset{\text{def}}{=} \max_i \|a_i - b_i\|.$$

Proposition. *Let K be a finite extension of $\mathbb{Q}_p$. Let $f(X) \in K[X]$ have degree n*

$$f(X) = a_n X^n + a_{n-1} X^{n-1} + \cdots + a_1 X + a_0.$$

Suppose the roots of f in $\overline{\mathbb{Q}}_p$ are distinct. Then for every $\varepsilon > 0$ there exists a δ such that, if $g = \sum_{i=0}^n b_i X^i \in K[X]$ has degree n, and if $|f - g|_p < \delta$, then for every root α_i of $f(X)$ there is precisely one root β_i of $g(X)$ such that $|\alpha_i - \beta_i|_p < \varepsilon$.

PROOF. For each root β of $g(X)$ we have

$$|f(\beta)|_p = |f(\beta) - g(\beta)|_p = \left| \sum_{i=0}^{n} (a_i - b_i)\beta^i \right|_p$$

$$\leq \max_i (|a_i - b_i|_p |\beta|_p^i)$$

$$\leq |f - g|_p \max(1, |\beta|_p^n) < \delta C_1^n,$$

where C_1 is a suitable constant (see Exercise 3 below).

Let $C_2 = \min_{1 \leq i < j \leq n} |\alpha_i - \alpha_j|_p$. Since the α_i's are distinct, we have $C_2 \neq 0$. Then the relation $|\beta - \alpha_i|_p < C_2$ is only possible for at most one α_i (since if it held for another $\alpha_j \neq \alpha_i$ we'd have $|\alpha_i - \alpha_j|_p \leq \max(|\alpha_i - \beta|_p, |\beta - \alpha_j|_p) < C_2$). Since

$$C_1^n \delta > |f(\beta)|_p$$

$$= \left| a_n \prod (\beta - \alpha_i) \right|_p \quad (\text{since } f(X) = a_n \prod (X - \alpha_i))$$

$$= |a_n|_p \prod |\beta - \alpha_i|_p,$$

it's clear that for δ sufficiently small such an α_i with $|\beta - \alpha_i|_p < C_2$ must exist. Moreover, for that α_i we have:

$$|\beta - \alpha_i|_p < \frac{C_1^n \delta}{|a_n|_p \prod_{j \neq i} |\beta - \alpha_j|_p}$$

$$\leq \frac{C_1^n \delta}{|a_n|_p \cdot C_2^{n-1}},$$

which can be made $< \varepsilon$ by a suitable choice of δ. $\qquad \square$

4. Ω

So far we've been dealing only with algebraic extensions of $\mathbb{Q}_p$. But, as mentioned before, this is not yet enough to give us the p-adic analogue of the complex numbers.

Theorem 12. $\bar{\mathbb{Q}}_p$ *is not complete.*

PROOF. We must give an example of a Cauchy sentence $\{a_i\}$ in $\bar{\mathbb{Q}}_p$ such that there cannot exist a number $a \in \bar{\mathbb{Q}}_p$ which is the limit of the a_i.

Let b_i be a primitive $(p^{2^i} - 1)$th root of 1 in $\bar{\mathbb{Q}}_p$, i.e., $b_i^{p^{2^i}-1} = 1$, but $b_i^m \neq 1$ if $m < p^{2^i} - 1$. Note that $b_i^{p^{2^{i'}}-1} = 1$ if $i' > i$, because $2^i | 2^{i'}$ implies $p^{2^i} - 1 | p^{2^{i'}} - 1$. (In fact, instead of 2^i we could replace the exponent of p by any increasing sequence whose ith term divides its $(i + 1)$th, e.g., 3^i, $i!$, etc.) Thus, if $i' > i$, b_i is a power of $b_{i'}$. Let

$$a_i = \sum_{j=0}^{i} b_j p^{N_j},$$

where $0 = N_0 < N_1 < N_2 < \cdots$ is an increasing sequence of nonnegative integers that will be chosen later. Note that the b_j, $j = 0, 1, \ldots, i$, are the digits in the p-adic expansion of a_i in the unramified extension $\mathbb{Q}_p(b_i)$, since the b_j are Teichmüller representatives. Clearly $\{a_i\}$ is Cauchy.

We now choose the N_j, $j > 0$, by induction. Suppose we have defined N_j for $j \leq i$, so that we have our $a_i = \sum_{j=0}^{i} b_j p^{N_j}$. Let $K = \mathbb{Q}_p(b_i)$. In §3 we proved that K is a Galois unramified extension of degree 2^i. First note that $\mathbb{Q}_p(a_i) = K$, because otherwise there would be a nontrivial $\mathbb{Q}_p$-automorphism σ of K which leaves a_i fixed (see paragraph (11) in §1). But $\sigma(a_i)$ has p-adic expansion $\sum_{j=0}^{i} \sigma(b_j) p^{N_j}$, and $\sigma(b_i) \neq b_i$, so that $\sigma(a_i) \neq a_1$ because they have different p-adic expansions.

Next, by exercise 9 of §III.1, there exists $N_{i+1} > N_i$ such that a_i does not satisfy any congruence

$$\alpha_n a_i^n + \alpha_{n-1} a_i^{n-1} + \cdots + \alpha_1 a_i + \alpha_0 \equiv 0 \pmod{p^{N_{i+1}}}$$

for $n < 2^i$ and $\alpha_j \in \mathbb{Z}_p$ not all divisible by p.

This gives us our sequence $\{a_i\}$.

Suppose that $a \in \bar{\mathbb{Q}}_p$ were a limit of $\{a_i\}$. Then a satisfies an equation

$$\alpha_n a^n + \alpha_{n-1} a^{n-1} + \cdots + \alpha_1 a + \alpha_0 = 0,$$

where we may assume that all of the $\alpha_i \in \mathbb{Z}_p$ and not all are divisible by p. Choose i so that $2^i > n$. Since $a \equiv a_i \pmod{p^{N_{i+1}}}$, we have

$$\alpha_n a_i^n + \alpha_{n-1} a_i^{n-1} + \cdots + \alpha_1 a_i + \alpha_0 \equiv 0 \pmod{p^{N_{i+1}}},$$

a contradiction. This proves the theorem. □

Note that we have actually proved that $\mathbb{Q}_p^{\text{unram}}$, not only $\overline{\mathbb{Q}}_p = \mathbb{Q}_p^{\text{alg cl}}$, is not complete.

So we now want to "fill in the holes," and define a new field Ω to be the *completion* of $\overline{\mathbb{Q}}_p$. Strictly speaking, this means looking at equivalence classes of Cauchy sequences of elements in $\overline{\mathbb{Q}}_p$ and proceeding in exactly the same way as how $\mathbb{Q}_p$ was constructed from $\mathbb{Q}$ (or how $\mathbb{R}$ was constructed from $\mathbb{Q}$, or how a completion can be constructed for *any* metric space). Intuitively speaking, we're creating a new field Ω by throwing in all numbers which are convergent infinite sums of numbers in $\overline{\mathbb{Q}}_p$, for example, of the type considered in the proof of Theorem 12.

Just as in going from $\mathbb{Q}$ to $\mathbb{Q}_p$, in going from $\overline{\mathbb{Q}}_p$ to Ω we can extend the norm $|\ |_p$ on $\overline{\mathbb{Q}}_p$ to a norm on Ω be defining $|x|_p = \lim_{i \to \infty} |x_i|_p$, where $\{x_i\}$ is a Cauchy sequence of elements in $\overline{\mathbb{Q}}_p$ that is in the equivalence class of x (see §I.4). As in going from $\mathbb{Q}$ to $\mathbb{Q}_p$, it is easy to see that if $x \neq 0$ this limit $|x|_p$ is actually equal to $|x_i|_p$ for i sufficiently large.

We also extend ord_p to Ω:

$$\text{ord}_p x = -\log_p |x|_p.$$

Let $A = \{x \in \Omega | |x|_p \leq 1\}$ be the "valuation ring" of Ω, let $M = \{x \in \Omega | |x|_p < 1\}$ be its maximal ideal, and let $A^\times = \{x \in \Omega | |x|_p = 1\} = A - M$ be the set of invertible elements of A. Suppose that $x \in A^\times$, i.e., $|x|_p = 1$. Since $\overline{\mathbb{Q}}_p$ is dense in Ω, we can find an algebraic x' such that $x - x' \in M$, i.e., $|x - x'|_p < 1$. Since then $|x'|_p = 1$, it follows that x' is integral over $\mathbb{Z}_p$, i.e., it satisfies a monic polynomial with coefficients in $\mathbb{Z}_p$. Reducing that polynomial modulo p, we find that the coset $x + M = x' + M$ is *algebraic over* $\mathbb{F}_p$, i.e., lies in some $\mathbb{F}_{p^f}$. Now let $\omega(x)$ be the $(p^f - 1)$th root of 1 which is the Teichmüller representative of $x + M \in \mathbb{F}_{p^f}$, and set $\langle x \rangle = x/\omega(x)$. Then $\langle x \rangle \in 1 + M$. In other words, any $x \in A^\times$ is the product of a root of unity $\omega(x)$ and an element $\langle x \rangle$ which is in the open unit disc about 1. (If $x \in \mathbb{Z}_p$ has first digit a_0, this simply says that x is the product of the Teichmüller representative of a_0 and an element of $1 + p\mathbb{Z}_p$.) Finally, an arbitrary nonzero $x \in \Omega$ can be written as a fractional power of p times an element $x_1 \in \Omega$ of absolute value 1. Namely, if $\text{ord}_p x = r = a/b$ (see Exercise 1 below), then let p^r denote any root in $\overline{\mathbb{Q}}_p$ of the polynomial $X^b - p^a$. Then $x = p^r x_1 = p^r \omega(x_1) \langle x_1 \rangle$ for some x_1 of norm 1. In other words, *any nonzero element of Ω is a product of a fractional power of p, a root of unity, and an element in the open unit disc about 1.*

The next theorem tells us that we are done: Ω will serve as the p-adic analogue of the complex numbers.

Theorem 13. *Ω is algebraically closed.*

Proof. Let: $f(X) = X^n + a_{n-1}X^{n-1} + \cdots + a_1 X + a_0$, $a_i \in \Omega$. We must show that $f(X)$ has a root in Ω. For each $i = 0, 1, \ldots, n - 1$, let $\{a_{ij}\}_j$ be a sequence of elements of $\bar{\mathbb{Q}}_p$ which converge to a_i. Let $g_j(X) = X^n + a_{n-1,j}X^{n-1} + \cdots + a_{1,j}X + a_{0,j}$. Let r_{ij} be the roots of $g_j(X)$ ($i = 1, 2, \ldots, n$). We claim that we can find i_j ($1 \leq i_j \leq n$) for $j = 1, 2, 3, \ldots$ such that the sequence $\{r_{i_j,j}\}$ is Cauchy. Namely, suppose we have $r_{i_j,j}$ and we want to find $r_{i_{j+1},j+1}$. Let $\delta_j = |g_j - g_{j+1}|_p = \max_i(|a_{i,j} - a_{i,j+1}|_p)$ (which approaches 0 as $j \to \infty$). Let $A_j = \max(1, |r_{i_j,j}|_p^n)$. Clearly there is a uniform constant A such that $A_j \leq A$ for all j (see Exercise 3 below). Then we have

$$\prod_i |r_{i_j,j} - r_{i,j+1}|_p = |g_{j+1}(r_{i_j,j})|_p$$

$$= |g_{j+1}(r_{i_j,j}) - g_j(r_{i_j,j})|_p$$

$$\leq \delta_j A.$$

Hence at least one of the $|r_{i_j,j} - r_{i,j+1}|_p$ on the left is $\leq \sqrt[n]{\delta_j A}$. Let $r_{i_{j+1},j+1}$ be any such $r_{i,j+1}$. Clearly this sequence of $r_{i_j,j}$ is Cauchy.

Now let $r = \lim_{j\to\infty} r_{i_j,j} \in \Omega$. Then $f(r) = \lim_{j\to\infty} f(r_{i_j,j}) = \lim_{j\to\infty} g_j(r_{i_j,j}) = 0$. $\quad\square$

Summarizing Chapters I and III, we can say that we have constructed Ω, which is the smallest field which contains $\mathbb{Q}$ and is both algebraically closed and complete with respect to $|\ |_p$. (Strictly speaking, this can be seen as follows: let Ω' be any such field; since Ω' is complete, it must contain a field isomorphic to the p-adic completion of $\mathbb{Q}$, which we can call $\mathbb{Q}_p$; then, since Ω' contains $\mathbb{Q}_p$ and is algebraically closed, it must contain a field isomorphic to the algebraic closure of $\mathbb{Q}_p$, which we can call $\bar{\mathbb{Q}}_p$; and, since Ω' contains $\bar{\mathbb{Q}}_p$ and is complete, it must contain a field isomorphic to the completion of $\bar{\mathbb{Q}}_p$, which we call Ω. Thus any field with these properties must contain a field isomorphic to Ω. The point is that both completion and algebraic closure are unique processes up to isomorphism.)

Actually, Ω should be denoted Ω_p, so as to remind us that everything we're doing depends on the prime number p we fixed at the start. But for brevity of notation we shall omit the subscript p.

The field Ω is a beautiful, gigantic realm, in which p-adic analysis lives.

EXERCISES

1. Prove that the possible values of $|\ |_p$ on $\bar{\mathbb{Q}}_p^{\times}$ is the set of all rational powers of p (in the positive real numbers). What about on Ω? Recall that we let the ord_p function extend to $\Omega^{\times}$ by defining $\mathrm{ord}_p x = -\log_p |x|_p$ (i.e., the power $1/p$ is raised to get $|x|_p$). What is the set of all possible values of ord_p on $\Omega^{\times}$? Now prove that $\bar{\mathbb{Q}}_p$ and Ω are *not* locally compact. This is one striking difference with $\mathbb{C}$, which is locally compact under the Archimedean metric (the usual definition of distance on the complex plane).

2. What happens if you define an "ellipse" in Ω to be the set of points the sum of whose distance from two fixed points $a, b \in \Omega$ is a fixed real number r? Show that this "ellipse" is either two disjoint circles, the intersection of two circles, or the empty set, depending on a, b, and r. What do you get if you define a "hyperbola" as $\{x \in \Omega \mid |x - a|_p - |x - b|_p = r\}$?

3. Let $g(X) = X^n + b_{n-1}X^{n-1} + \cdots + b_1 X + b_0$. Let $C_0 \underset{\text{def}}{=} \max_i |b_i|_p$. Show that there exists a constant C_1 depending only on C_0 such that any root β of $g(X)$ satisfies $|\beta|_p < C_1$.

4. Let α be a root of a monic irreducible polynomial $f(X) \in K[X]$, where K is a finite extension of $\mathbb{Q}_p$. Prove that there exists an $\varepsilon > 0$ such that any polynomial $g(X)$ having the same degree as f and satisfying $|f - g|_p < \varepsilon$ has a root β such that $K(\alpha) = K(\beta)$. Show that this is not necessarily the case if f is not irreducible.

5. Prove that any finite extension K of $\mathbb{Q}_p$ contains a finite extension F of the rational numbers $\mathbb{Q}$ such that $[F : \mathbb{Q}] = [K : \mathbb{Q}_p]$ and F is *dense* in K, i.e., for any element $x \in K$ and any $\varepsilon > 0$ there exists $y \in F$ such that $|x - y|_p < \varepsilon$.

6. Let p be a prime such that -1 does not have a square root in $\mathbb{Q}_p$ (see Exercise 8 of §III.1). Use Krasner's Lemma to find an ε such that $\mathbb{Q}_p(\sqrt{-a}) = \mathbb{Q}_p(\sqrt{-1})$ whenever $|a - 1|_p < \varepsilon$. For what ε does $|a - p|_p < \varepsilon$ imply $\mathbb{Q}_p(\sqrt{a}) = \mathbb{Q}_p(\sqrt{p})$? (Treat the case $p = 2$ separately.)

7. Let a be a primitive p^nth root of 1 in $\overline{\mathbb{Q}}_p$, i.e., $a^{p^{n-1}} \neq 1$. Find $|a - 1|_p$. In the case $n = 1$ show that $\mathbb{Q}_p(a) = \mathbb{Q}_p(\beta)$ where β is any root in $\overline{\mathbb{Q}}_p$ of $X^{p-1} + p = 0$. Also show that $|a - 1|_p = 1$ if a is a primitive mth root of 1 and m is not a power of p.

8. Let K be a finite extension of $\mathbb{Q}_p$. Let m be a positive integer, and let $(K^\times)^m$ denote the set of all mth powers of elements of $K^\times$. Suppose that (1) $|m|_p = 1$, and (2) K contains no mth roots of 1 other than 1 itself. (For example, if $K = \mathbb{Q}_p$, these two conditions both hold if and only if m is relatively prime to both p and $p - 1$, as you can prove as an exercise.) Prove that the index of $(K^\times)^m$ as a multiplicative subgroup in $K^\times$ (i.e., the number of distinct cosets) is equal to m.

9. If in the previous exercise we remove the assumption that K contains no nontrivial m-th roots of 1, show that the index of $(K^\times)^m$ in $K^\times$ equals mw, where w is the number of m-th roots of 1 contained in K.

10. If K is a totally ramified extension of $\mathbb{Q}_p$, show that every mth root of 1 in K is in $\mathbb{Q}_p$ if p does not divide m.

11. Determine the cardinality of the sets $\mathbb{Q}_p$, $\overline{\mathbb{Q}}_p$, and Ω.

12. Prove that $ef = n$, where $n = [K : \mathbb{Q}_p]$, e is the index of ramification, and f is the residue field degree.

13. Let K be a totally ramified extension of $\mathbb{Q}_p$ of degree e. Show that there exists $\beta \in K$ such that $|\beta^e - \alpha|_p < 1/p$ for some $\alpha \in \mathbb{Z}_p$ with $\mathrm{ord}_p \alpha = 1$.

14. Suppose K is *tamely* totally ramified. Using a Hensel's lemma type argument, show that β can be further adjusted so that $\beta^e \in \mathbb{Q}_p$, i.e., β satisfies $X^e - \alpha = 0$, where $\alpha \in \mathbb{Z}_p$ and $\mathrm{ord}_p \alpha = 1$. Note that $K = \mathbb{Q}_p(\beta)$ (explain why).

15. The complex numbers $\mathbb{C}$ are much more numerous than the rational numbers, or even the algebraic numbers, because the latter sets are only countably

infinite, while $\mathbb{C}$ has the cardinality of the continuum. Ω is also much, much bigger than $\mathbb{Q}_p^{\text{alg cl}}$, although not in precisely that way (see Exercise 11 above). Prove that there does not exist a countably infinite set of elements of Ω such that Ω is an algebraic extension of the field obtained by adjoining all those elements to $\overline{\mathbb{Q}}_p$ (i.e., the field of all rational expressions involving those elements and elements of $\overline{\mathbb{Q}}_p$). One says that Ω has "uncountably infinite *transcendence degree* over $\overline{\mathbb{Q}}_p$." (*Warning*: this exercise and the next are hard!)

16. Does Ω have countably infinite transcendence degree over the p-adic completion of $\mathbb{Q}_p^{\text{unram}}$?

CHAPTER IV

p-adic power series

1. Elementary functions

Recall that in a metric space whose metric comes from a non-Archimedean norm $\| \ \|$, a sequence is Cauchy if and only if the difference between adjacent terms approaches zero; and if the metric space is complete, an infinite sum converges if and only if its general term approaches zero. So if we consider expressions of the form

$$f(X) = \sum_{n=0}^{\infty} a_n X^n, \qquad a_n \in \Omega,$$

we can give a value $\sum_{n=0}^{\infty} a_n x^n$ to $f(x)$ whenever an x is substituted for X for which $|a_n x^n|_p \to 0$.

Just as in the Archimedean case (power series over $\mathbb{R}$ or $\mathbb{C}$), we define the "radius of convergence"

$$r = \frac{1}{\lim \sup |a_n|_p^{1/n}},$$

where the terminology "$1/r = \lim \sup |a_n|_p^{1/n}$" means that $1/r$ is the *least* real number such that for *any* $C > 1/r$ there are only finitely many $|a_n|_p^{1/n}$ greater than C. Equivalently, $1/r$ is the greatest "point of accumulation," i.e., the greatest real number which can occur as the limit of a subsequence of $\{|a_n|_p^{1/n}\}$. If, for example, $\lim_{n \to \infty} |a_n|_p^{1/n}$ exists, then $1/r$ is simply this limit.

We justify the use of the term "radius of convergence" by showing that the series converges if $|x|_p < r$ and diverges if $|x|_p > r$. First, if $|x|_p < r$, then, letting $|x|_p = (1 - \varepsilon)r$, we have: $|a_n x^n|_p = (r|a_n|_p^{1/n})^n (1 - \varepsilon)^n$. Since there are only finitely many n for which $|a_n|_p^{1/n} > 1/(r - \tfrac{1}{2}\varepsilon r)$, we have

$$\lim_{n \to \infty} |a_n x^n|_p \leq \lim_{n \to \infty} \left(\frac{(1 - \varepsilon)r}{(1 - \tfrac{1}{2}\varepsilon)r} \right)^n = \lim_{n \to \infty} \left(\frac{1 - \varepsilon}{1 - \tfrac{1}{2}\varepsilon} \right)^n = 0.$$

Similarly, we easily see that if $|x|_p > r$, then $a_n x^n$ does not approach 0 as $n \to \infty$. Finally, if $r = \infty$ it is easy to check that $\lim_{n \to \infty} |a_n x^n|_p = 0$ for all x.

What if $|x|_p = r$? In the Archimedean case the story on the boundary of the interval or disc of convergence can be a little complicated. For example, $\log(1 + x) = \sum_{n=1}^{\infty} (-1)^{n+1} x^n/n$ has radius of convergence 1. When $|x| = 1$, it diverges for $x = -1$ and converges ("conditionally," not "absolutely") for other values of x (i.e., for $x = 1$ in the case of the reals and on the unit circle minus the point $x = -1$ in the case of the complexes).

But in the non-Archimedean case there's a single answer for all points $|x|_p = r$. This is because a series converges if and only if its terms approach zero, i.e., if and only if $|a_n|_p |x|_p^n \to 0$, and this depends only on the norm $|x|_p$ and not on the particular value of x with a given norm—there's no such thing as "conditional" convergence ($\sum \pm a_n$ converging or diverging depending on the choices of $\pm$'s).

If we take the same example $\sum_{n=1}^{\infty} (-1)^{n+1} X^n/n$, we find that $|a_n|_p = p^{\operatorname{ord}_p n}$, and $\lim_{n \to \infty} |a_n|_p^{1/n} = 1$. The series converges for $|x|_p < 1$ and diverges for $|x|_p > 1$. If $|x|_p = 1$, then $|a_n x^n|_p = p^{\operatorname{ord}_p n} \geq 1$, and the series diverges for all such x.

Now let's introduce some notation. If R is a ring, we let $R[[X]]$ be the ring of formal power series in X with coefficients in R, i.e., expressions $\sum_{n=0}^{\infty} a_n X^n$, $a_n \in R$, which add and multiply together in the usual way. For us, R will usually be $\mathbb{Z}$, $\mathbb{Q}$, $\mathbb{Z}_p$, $\mathbb{Q}_p$, or Ω. We often abbreviate other sets using this notation, for example,

$$1 + XR[[X]] \underset{\text{def}}{=} \{f \in R[[X]] \mid \text{constant term } a_0 \text{ of } f \text{ is } 1\}.$$

We define the "closed disc of radius $r \in \mathbb{R}$ about a point $a \in \Omega$" to be

$$D_a(r) \underset{\text{def}}{=} \{x \in \Omega \mid |x - a|_p \leq r\},$$

and we define the "open disc of radius r about a" to be

$$D_a(r^-) \underset{\text{def}}{=} \{x \in \Omega \mid |x - a|_p < r\}.$$

We let $D(r) \underset{\text{def}}{=} D_0(r)$ and $D(r^-) \underset{\text{def}}{=} D_0(r^-)$. (Note: whenever we refer to the closed disc $D(r)$ in Ω, we understand r to be a possible value of $|\ |_p$, i.e., a positive real number that is a rational power of p; we always write $D(r^-)$ if there are no $x \in \Omega$ with $|x|_p = r$.)

(A word of caution. The terms "closed" and "open" are used only out of analogy with the Archimedean case. From a topological point of view the terminology is bad. Namely, for positive c the set $C_c = \{x \in \Omega \mid |x - a|_p = c\}$ is open in the topological sense, because every point $x \in C_c$ has a disc about it, for example $D_x(c^-)$, all points of which belong to C_c. But then any union of C_c's is open. Both $D_a(r)$ and $D_a(r^-)$, as well as their complements, are such unions: for example, $D_a(r^-) = \bigcup_{c < r} C_c$. Hence both $D_a(r)$ and $D_a(r^-)$ are simultaneously *open and closed* sets. The term for this peculiar state of affairs in Ω is "totally disconnected topological space.")

Just to get used to the notation, we prove a trivial lemma.

Lemma 1. *Every $f(X) \in \mathbb{Z}_p[[X]]$ converges in $D(1^-)$.*

PROOF. Let $f(X) = \sum_{n=0}^{\infty} a_n X^n$, $a_n \in \mathbb{Z}_p$, and let $x \in D(1^-)$. Thus, $|x|_p < 1$. Also $|a_n|_p \leq 1$ for all n. Hence $|a_n x^n|_p \leq |x|_p^n \to 0$ as $n \to \infty$. $\qquad\square$

Another easy lemma is

Lemma 2. *Every $f(X) = \sum_{n=0}^{\infty} a_n X^n \in \Omega[[X]]$ which converges in an (open or closed) disc $D = D(r)$ or $D(r^-)$ is continuous on D.*

PROOF. Suppose $|x' - x|_p < \delta$, where $\delta < |x|_p$ will be chosen later. Then $|x'|_p = |x|_p$. (We are assuming $x \neq 0$; the case $x = 0$ is very easy to check separately.) We have

$$|f(x) - f(x')|_p = \left| \sum_{n=0}^{\infty} (a_n x^n - a_n x'^n) \right|_p$$

$$\leq \max_n |a_n x^n - a_n x'^n|_p$$

$$= \max_n (|a_n|_p |(x - x')(x^{n-1} + x^{n-2}x' + \cdots$$

$$+ xx'^{n-2} + x'^{n-1}|_p).$$

But $|x^{n-1} + x^{n-2}x' + \cdots + xx'^{n-2} + x'^{n-1}|_p \leq \max_{1 \leq i \leq n} |x^{n-i}x'^{i-1}|_p = |x|_p^{n-1}$. Hence

$$|f(x) - f(x')|_p \leq \max_n (|x - x'|_p |a_n|_p |x|_p^{n-1})$$

$$< \frac{\delta}{|x|_p} \max_n (|a_n|_p |x|_p^n).$$

Since $|a_n|_p |x|_p^n$ is bounded as $n \to \infty$, this $|f(x) - f(x')|_p$ is $< \varepsilon$ for suitable δ. $\qquad\square$

Now let's return to our series $\sum_{n=1}^{\infty} (-1)^{n+1} X^n / n$, which, as we've seen, has disc of convergence $D(1^-)$. That is, this series gives a function on $D(1^-)$ taking values in Ω. Let's call this function $\log_p(1 + X)$, where the subscript p reminds us of the prime which gave us the norm on $\mathbb{Q}$ used to get Ω, and also reminds us not to confuse this function with the classical $\log(1 + X)$ function—which has a different domain (a subset of $\mathbb{R}$ or $\mathbb{C}$) and range ($\mathbb{R}$ or $\mathbb{C}$). Unfortunately, the notation $\log_p$ for the "p-adic logarithm" is identical to classical notation for "log to the base p." From now on, we shall assume that $\log_p$ means p-adic logarithm

$$\log_p(1 + X): D(1^-) \to \Omega, \qquad \log_p(1 + x) = \sum_{n=1}^{\infty} (-1)^{n+1} x^n / n,$$

unless *explicitly* stated otherwise.

The dangers of confusing Archimedean and p-adic functions will be illustrated below, and also in Exercises 8–10 at the end of §1.

Anyone who has studied differential equations (and many who haven't) realize that $\exp(x) = e^x = \sum_{n=0}^{\infty} x^n/n!$ is about the most important function there is in classical mathematics. So let's look at the series $\sum_{n=0}^{\infty} X^n/n!$ p-adically. The classical exponential series converges everywhere, thanks to the $n!$ in the denominator. But while big denominators are good things to have classically, they are not so good p-adically. Namely, it's not hard to compute (see Exercise 14 §I.2)

$$\text{ord}_p(n!) = \frac{n - S_n}{p - 1} \quad (S_n = \text{sum of digits in } n \text{ to base } p);$$

$$|1/n!|_p = p^{(n - S_n)/(p - 1)}.$$

Our formula for the radius of convergence $r = 1/(\lim \sup |a_n|_p^{1/n})$ gives us

$$\text{ord}_p r = \lim \inf\left(\frac{1}{n} \text{ord}_p a_n\right),$$

(where the "lim inf" of a sequence is its smallest point of accumulation). In the case $a_n = 1/n!$, this gives

$$\text{ord}_p r = \lim \inf\left(-\frac{n - S_n}{n(p - 1)}\right);$$

but $\lim_{n \to \infty}(-(n - S_n)/(n(p - 1))) = -1/(p - 1)$. Hence $\sum_{n=0}^{\infty} x^n/n!$ 'converges if $|x|_p < p^{-1/(p-1)}$ and diverges if $|x|_p > p^{-1/(p-1)}$. What if $|x|_p = p^{-1/(p-1)}$, i.e., $\text{ord}_p x = 1/(p - 1)$? In that case

$$\text{ord}_p(a_n x^n) = -\frac{n - S_n}{p - 1} + \frac{n}{p - 1} = \frac{S_n}{p - 1}.$$

If, say, we choose $n = p^m$ to be a power of p, so that $S_n = 1$, we have: $\text{ord}_p(a_{p^m} x^{p^m}) = 1/(p - 1)$, $|a_{p^m} x^{p^m}|_p = p^{-1/(p-1)}$, and hence $a_n x^n \not\to 0$ as $n \to \infty$. Thus, $\sum_{n=0}^{\infty} X^n/n!$ has disc of convergence $D(p^{-1/(p-1)-})$ (the $^-$ denoting the open disc, as usual). Let's denote $\exp_p(X) \underset{\text{def}}{=} \sum_{n=0}^{\infty} X^n/n! \in \mathbb{Q}_p[[X]]$.

Note that $D(p^{-1/(p-1)-}) \subset D(1^-)$, so that $\exp_p$ converges in a *smaller* disc than $\log_p$!

While it is important to avoid confusion between log and exp and $\log_p$ and $\exp_p$, we can carry over some basic properties of log and exp to the p-adic case. For example, let's try to get the basic property of log that log of a product equals the sum of the logs. Note that if $x \in D(1^-)$ and $y \in D(1^-)$, then also $(1 + x)(1 + y) = 1 + (x + y + xy) \in 1 + D(1^-)$. Thus, we have:

$$\log_p[(1 + x)(1 + y)] = \sum_{n=1}^{\infty} (-1)^{n+1}(x + y + xy)^n/n.$$

Meanwhile, we have the following relation in the ring of power series over $\mathbb{Q}$ in *two* indeterminates (written $\mathbb{Q}[[X, Y]]$):

$$\sum (-1)^{n+1} X^n/n + \sum (-1)^{n+1} Y^n/n = \sum (-1)^{n+1}(X + Y + XY)^n/n.$$

This holds because over $\mathbb{R}$ or $\mathbb{C}$ we have $\log(1 + x)(1 + y) = \log(1 + x) + \log(1 + y)$, so that the difference between the two sides of the above equality—call it $F(X, Y)$—must vanish for all real values of X and Y in the interval $(-1, 1)$. So the coefficient of $X^m Y^n$ in $F(X, Y)$ must vanish for all m and n.

The argument for why $F(X, Y)$ vanishes as a formal power series is typical of a line of reasoning we shall often need. Suppose that an expression involving some power series in X and Y—e.g., $\log(1 + X)$, $\log(1 + Y)$, and $\log(1 + X + Y + XY)$—vanishes whenever real values in some interval are substituted for the variables. Then when we gather together all $X^m Y^n$-terms in this expression, its coefficient must always be zero. Since this is a general fact unrelated to p-adic numbers, we won't digress to prove it carefully here. But if you have any doubts about whether you could prove this fact, turn to Exercise 21 below for further explanations and hints on how to prove it.

Returning to the p-adic situation, we note that if a series converges in Ω, its terms can be rearranged in any order, and the resulting series converges to the same limit. (This is easy to·check—it's related to there being no such thing as "conditional" convergence.) Thus, $\log_p[(1 + x)(1 + y)] = \sum_{n=1}^{\infty} (-1)^{n+1}(x + y + xy)^n/n$ can be written as $\sum_{m,n=0}^{\infty} c_{m,n}x^n y^m$. But the "formal identity" in $\mathbb{Q}[[X, Y]]$ tells us that the rational numbers $c_{m,n}$ will be 0 unless $n = 0$ or $m = 0$, in which case: $c_{0,n} = c_{n,0} = (-1)^{n+1}/n$ ($c_{0,0} = 0$). In other words, we may conclude that

$$\log_p[(1 + x)(1 + y)] = \sum_{n=1}^{\infty} (-1)^{n+1}x^n/n + \sum_{n=1}^{\infty} (-1)^{n+1}y^n/n$$
$$= \log_p(1 + x) + \log_p(1 + y).$$

As a corollary of this formula, take the case when $1 + x$ is a p^mth root of 1. Then $|x|_p < 1$ (see Exercise 7 of §III.4), so that: $p^m \log_p(1 + x) = \log_p (1 + x)^{p^m} = \log_p 1 = 0$. Hence $\log_p(1 + x) = 0$.

In exactly the same way we can prove the familiar rule for exp in the p-adic situation: if $x, y \in D(p^{-1/(p-1)-})$, then $x + y \in D(p^{-1/(p-1)-})$, and $\exp_p(x + y) = \exp_p x \cdot \exp_p y$.

Moreover, we also find a result analogous to the Archimedean case as far as $\log_p$ and $\exp_p$ being inverse functions of one another. More precisely, suppose $x \in D(p^{-1/(p-1)-})$. Then $\exp_p x = 1 + \sum_{n=1}^{\infty} x^n/n!$, and $\mathrm{ord}_p(x^n/n!) > n/(p - 1) - (n - S_n)/(p - 1) = S_n/(p - 1) > 0$. Thus, $\exp_p x - 1 \in D(1^-)$. Suppose we take

$$\log_p(1 + \exp_p x - 1) = \sum_{n=1}^{\infty} (-1)^{n+1}(\exp_p x - 1)^n/n$$
$$= \sum_{n=1}^{\infty} (-1)^{n+1} \left(\sum_{m=1}^{\infty} x^m/m! \right)^n \Big/ n.$$

But this series can be rearranged to get a series of the form $\sum_{n=1}^{\infty} c_n x^n$. And reasoning as before, we have the following formal identity over $\mathbb{Q}[[X, Y]]$:

$$\sum_{n=1}^{\infty} (-1)^{n+1} \left(\sum_{m=1}^{\infty} X^m/m! \right)^n \Big/ n = X,$$

coming from the fact that $\log(\exp x) = x$ over $\mathbb{R}$ or $\mathbb{C}$. Hence $c_1 = 1$, $c_n = 0$ for $n > 1$, and

$$\log_p(1 + \exp_p x - 1) = x \quad \text{for } x \in D(p^{-1/(p-1)-}).$$

To go the other way—i.e., $\exp_p(\log_p(1 + x))$—we have to be a little careful, because even if x is in the region of convergence $D(1^-)$ of $\log_p(1 + X)$, it is *not* necessarily the case that $\log_p(1 + x)$ is in the region of convergence $D(p^{-1/(p-1)-})$ of $\exp_p X$. This *is* the case if $x \in D(p^{-1/(p-1)-})$, since then for $n \geq 1$:

$$(\text{ord}_p\, x^n/n) - \frac{1}{p-1} > \frac{n}{p-1} - \text{ord}_p\, n - \frac{1}{p-1} = \frac{n-1}{p-1} - \text{ord}_p\, n,$$

which has its minima at $n = 1$ and $n = p$, where it's zero. Thus, $\text{ord}_p \log_p (1 + x) \geq \min_n \text{ord}_p\, x^n/n > 1/(p-1)$. Then everything goes through as before, and we have:

$$\exp_p(\log_p(1 + x)) = 1 + x \quad \text{for } x \in D(p^{-1/(p-1)-}).$$

All of the facts we have proved about $\log_p$ and $\exp_p$ can be stated succinctly in the following way.

Proposition. *The functions* $\log_p$ *and* $\exp_p$ *give mutually inverse isomorphisms between the multiplicative group of the open disc of radius* $p^{-1/(p-1)}$ *about* 1 *and the additive group of the open disc of radius* $p^{-1/(p-1)}$ *about* 0. (This means precisely the following: $\log_p$ gives a one-to-one correspondence between the two sets, under which the image of the product of two numbers is the sum of the images, and $\exp_p$ is the inverse map.)

This isomorphism is analogous to the real case, where log and exp give mutually inverse isomorphisms between the multiplicative group of positive real numbers and the additive group of all real numbers.

In particular, this proposition says that $\log_p$ is injective on $D_1(p^{-1/(p-1)-})$, i.e., no two numbers in $D_1(p^{-1/(p-1)-})$ have the same $\log_p$. It's easy to see that $D_1(p^{-1/(p-1)-})$ is the biggest disc on which this is true: namely, a primitive pth root ζ of 1 has $|\zeta - 1|_p = p^{-1/(p-1)}$ (see Exercise 7 of §III.4), and also $\log_p \zeta = 0 = \log_p 1$.

We can similarly define the functions

$$\sin_p: D(p^{-1/(p-1)-}) \to \Omega, \qquad \sin_p X = \sum_{n=0}^{\infty} (-1)^n X^{2n+1}/(2n+1)!;$$

$$\cos_p: D(p^{-1/(p-1)-}) \to \Omega, \qquad \cos_p X = \sum_{n=0}^{\infty} (-1)^n X^{2n}/(2n)!.$$

Another function which is important in classical mathematics is the binomial expansion $B_a(x) = (1 + x)^a = \sum_{n=0}^{\infty} a(a-1) \cdots (a-n+1)/n!\, x^n$. For any $a \in \mathbb{R}$ or $\mathbb{C}$, this series converges in $\mathbb{R}$ or $\mathbb{C}$ if $|x| < 1$ and diverges if $|x| > 1$ (unless a is a nonnegative integer); its behavior at $|x| = 1$ is a little complicated, and depends on the value of a.

Now for any $a \in \Omega$ let's define

$$B_{a,p}(X) \underset{\text{def}}{=} \sum_{n=0}^{\infty} \frac{a(a-1)\cdots(a-n+1)}{n!} X^n,$$

and proceed to investigate its convergence. First of all, suppose $|a|_p > 1$. Then $|a - i|_p = |a|_p$, and the nth term has $|\ |_p$ equal to $|ax|_p^n/|n!|_p$. Thus, for $|a|_p > 1$, the series $B_{a,p}(X)$ has region of convergence $D((p^{-1/(p-1)})/|a|_p-)$.

Now suppose $|a|_p \le 1$. The picture becomes more complicated, and depends on a. We won't derive a complete answer. In any case, for any such a we have $|a - i|_p \le 1$, and so $|a(a-1)\cdots(a-n+1)/n!\, x^n|_p \le |x^n/n!|_p$, so that at least $B_{a,p}(X)$ converges on $D(p^{-1/(p-1)}-)$.

We'll soon need a more accurate result about the convergence of $B_{a,p}(X)$ in the case when $a \in \mathbb{Z}_p$. We claim that then $B_{a,p}(X) \in \mathbb{Z}_p[[X]]$ (and, in particular, it converges on $D(1^-)$ by Lemma 1). Thus, we want to show that $a(a-1)\cdots(a-n+1)/n! \in \mathbb{Z}_p$. Let a_0 be a positive integer greater than n such that $\text{ord}_p(a - a_0) > N$ (N will be chosen later). Then $a_0(a_0-1)\cdots(a_0-n+1)/n! = \binom{a_0}{n} \in \mathbb{Z} \subset \mathbb{Z}_p$. It now suffices to show that for suitable N the difference between $a_0(a_0-1)\cdots(a_0-n+1)/n!$ and $a(a-1)\cdots(a-n+1)/n!$ has $|\ |_p \le 1$. But this follows because the polynomial $X(X-1)\cdots(X-n+1)$ is continuous. Thus,

$$B_{a,p}(X) \in \mathbb{Z}_p[[X]] \text{ if } a \in \mathbb{Z}_p.$$

As an important example of the case $a \in \mathbb{Z}_p$, suppose that $a = 1/m$, $m \in \mathbb{Z}$, $p \nmid m$. Let $x \in D(1^-)$. Then it follows by the same argument as used to prove $\log_p(1 + x)(1 + y) = \log_p(1 + x) + \log_p(1 + y)$ that we have

$$[B_{1/m,p}(x)]^m = 1 + x.$$

Thus, $B_{1/m,p}(x)$ is an mth root of $1 + x$ in Ω. (If $p|m$, this still holds, but now we can only substitute values of x in $D(|m|_p p^{-1/(p-1)}-)$.) So, whenever a is an ordinary rational number we can adopt the shorthand: $B_{a,p}(X) = (1 + X)^a$.

But be careful! What about the following "paradox"? Consider $4/3 = (1 + 7/9)^{1/2}$; in $\mathbb{Z}_7$ we have $\text{ord}_7 7/9 = 1$, and so for $x = 7/9$ and $n \ge 1$:

$$\left| \frac{1/2(1/2 - 1)\cdots(1/2 - n + 1)}{n!} x^n \right|_7 \le 7^{-n}/|n!|_7 < 1.$$

Hence

$$1 > |(1 + \tfrac{7}{9})^{1/2} - 1|_7 = |\tfrac{4}{3} - 1|_7 = |\tfrac{1}{3}|_7 = 1.$$

What's wrong??

Well, we were sloppy when we wrote $4/3 = (1 + 7/9)^{1/2}$. In both $\mathbb{R}$ and $\mathbb{Q}_7$ the number $16/9$ has two square roots $\pm 4/3$. In $\mathbb{R}$, the series for $(1 + 7/9)^{1/2}$ converges to $4/3$, i.e., the positive value is favored. But in $\mathbb{Q}_7$, the square root congruent to $1 \bmod 7$, i.e., $-4/3 = 1 - 7/3$, is favored. Thus, the *exact same series* of rational numbers

$$\sum_{n=0}^{\infty} \frac{1/2(1/2 - 1)\cdots(1/2 - n + 1)}{n!} \left(\frac{7}{9}\right)^n$$

converges to a rational number both 7-adically and in the Archimedean absolute value; but the rational numbers it converges to are different! This is a counterexample to the following false "theorem."

Non-theorem 1. *Let $\sum_{n=1}^{\infty} a_n$ be a sum of rational numbers which converges to a rational number in $|\ |_p$ and also converges to a rational number in $|\ |_\infty$. Then the rational value of the infinite sum is the same in both metrics.*

For more "paradoxes," see Exercises 8–10.

EXERCISES

1. Find the exact disc of convergence (specifying whether open or closed) of the following series. In (v) and (vi), $\log_p$ means the old-fashioned log to base p, and in (vii) ζ is a primitive pth root of 1. [ ] means the greatest integer function.

   (i) $\sum n!\, X^n$ (iii) $\sum p^n X^n$ (v) $\sum p^{[\log_p n]} X^n$ (vii) $\sum (\zeta - 1)^n X^n/n!$

   (ii) $\sum p^{n[\log n]} X^n$ (iv) $\sum p^n X^{p^n}$ (vi) $\sum p^{[\log_p n]} X^n/n$

2. Prove that, if $\sum a_n$ and $\sum b_n$ converge to a and b, respectively (where a_i, b_i, a, $b \in \Omega$), then $\sum c_n$, where $c_n = \sum_{i=0}^{n} a_i b_{n-i}$, converges to ab.

3. Prove that $1 + X\mathbb{Z}_p[[X]]$ is a group with respect to multiplication. Let D be an open or a closed disc in Ω of some radius about 0. Prove that $\{f \in 1 + X\Omega[[X]] \mid f$ converges on $D\}$ is closed under multiplication, but is not a group. Prove that for fixed λ, the set of $f(X) = 1 + \sum_{i=1}^{\infty} a_i X^i$ such that $\operatorname{ord}_p a_i - \lambda i$ is greater than 0 for all $i = 1, 2, \ldots$ and approaches ∞ as $i \to \infty$, is a multiplicative group. Next, let $f_j \in 1 + X\mathbb{Z}_p[[X]]$, $j = 1, 2, 3, \ldots$. Let $f(X) = \prod_{j=1}^{\infty} f_j(X^j)$. Check that $f(X) \in 1 + X\mathbb{Z}_p[[X]]$. Suppose that all of the f_j converge in the closed unit disc $D(1)$. Does $f(X)$ converge in $D(1)$ (proof or counterexample)? If all of the nonconstant coefficients of all of the f_j are divisible by p, does that change your answer (proof or counterexample)?

4. Let $\{a_n\} \subset \Omega$ be a sequence with $|a_n|_p$ bounded. Prove that

$$\sum_{n=0}^{\infty} a_n \frac{n!}{x(x+1)(x+2)\cdots(x+n)}$$

converges for all $x \in \Omega$ not in $\mathbb{Z}_p$. What can you say if $x \in \mathbb{Z}_p$?

5. Let i be a square root of -1 in $\bar{\mathbb{Q}}_p$ (actually, i lies in $\mathbb{Q}_p$ itself unless $p \equiv 3$ mod 4). Prove that: $\exp_p(ix) = \cos_p x + i \sin_p x$ for $x \in D(p^{-1/(p-1)-})$.

6. Show that $2^{p-1} \equiv 1 \pmod{p^2}$ if and only if p divides $\sum_{j=1}^{p-1} (-1)^j/j$ (of course, meaning that p divides the numerator of this fraction).

7. Show that the 2-adic ordinal of the rational number

$$2 + 2^2/2 + 2^3/3 + 2^4/4 + 2^5/5 + \cdots + 2^n/n$$

approaches infinity as n increases. Get a good estimate for this 2-adic ordinal in terms of n. Can you think of an entirely elementary proof (i.e., without using p-adic analysis) of this fact, which is actually completely elementary in its statement?

8. Find the fallacy in the following too-good-to-be-true proof of the irrationality of π. Suppose $\pi = a/b$. Let $p \neq 2$ be a prime not dividing a. Then

$$0 = \sin(pb\pi) = \sin(pa) = \sum_{n=0}^{\infty} (-1)^n (pa)^{2n+1}/(2n+1)! \equiv pa \pmod{p^2},$$

which is absurd.

9. Find the fallacy in the following proof of the transcendence of e. Suppose e were algebraic. Then $e - 1$ would also be algebraic. Choose a prime $p \neq 2$ which does not divide either the numerator or denominator of any coefficient of the monic irreducible polynomials satisfied by e and by $e - 1$ over $\mathbb{Q}$. You can show as an exercise that this implies that $|e|_p = |e - 1|_p = 1$. We have: $1 = |e - 1|_p^p = |(e - 1)^p|_p = |e^p - 1 - \sum_{i=1}^{p-1}\binom{p}{i}(-e)^i|_p$. Since the binomial coefficients in the summation are all divisible by p, and since $|-e|_p = 1$, it follows that $1 = |e^p - 1|_p = |\sum_{n=1}^{\infty} p^n/n!|_p$, which is impossible since each summand has $|\ |_p < 1$.

10. (*a*) Show that the binomial series for $(1 - p/(p + 1))^{-n}$ (where n is a positive rational integer) and for $(1 + (p^2 + 2mp)/m^2)^{1/2}$ (where m is a rational integer with $m > (\sqrt{2} + 1)p$, $p \nmid m$) converge to the same rational number as real and as p-adic infinite sums.
(*b*) Let $p \geq 7$, $n = (p - 1)/2$. Show that $(1 + p/n^2)^{1/2}$ gives a counter-example to Non-theorem 1.

11. Suppose that $\alpha \in \mathbb{Q}$ is such that $1 + \alpha$ is the square of a nonzero rational number a/b (written in lowest terms, with a and b positive). Let S be the set of all primes p for which the binomial series for $(1 + \alpha)^{1/2}$ converges in $|\ |_p$. Thus, $p \in S$ implies that $(1 + \alpha)^{1/2}$ converges to either a/b or $-a/b$ in $|\ |_p$. We also include the "infinite prime" in S if the binomial series converges in $|\ |_\infty = |\ |$, i.e., if $\alpha \in (-1, 1)$. Prove that:

(a) For p an odd prime, $p \in S$ if and only if $p|a + b$ or $p|a - b$, in which case $(1 + \alpha)^{1/2}$ converges to $-a/b$ when $p|a + b$ and to a/b when $p|a - b$.
(b) $2 \in S$ if and only if both a and b are odd, in which case $(1 + \alpha)^{1/2}$ converges to a/b when $a \equiv b \pmod 4$ and to $-a/b$ when $a \equiv -b \pmod 4$.
(c) $\infty \in S$ if and only if $0 < a/b < \sqrt{2}$, in which case $(1 + \alpha)^{1/2}$ converges to a/b.
(d) There is no α for which S is the empty set, and S consists of one element if $\alpha = 8$, $\frac{16}{9}$, 3, $\frac{5}{4}$ and for no other α.
(e) There is no α other than $8, \frac{16}{9}, 3, \frac{5}{4}$ for which $(1 + \alpha)^{1/2}$ converges to the same value in $|\ |_p$ for all $p \in S$. (This is one example of a very general theory of E. Bombieri.)

12. Prove that for any nonnegative integer k, the p-adic number $\sum_{n=0}^{\infty} n^k p^n$ is in $\mathbb{Q}$.

13. Prove that in $\mathbb{Q}_3$:

$$\sum_{n=1}^{\infty} (-1)^n \frac{3^{2n}}{n 4^{2n}} = 2 \cdot \sum_{n=1}^{\infty} \frac{3^{2n}}{n 4^n}.$$

14. Show that the disc of convergence of a power series $f(X) = \sum a_n X^n$ is contained in the disc of convergence of its derivative power series $f'(X) = \sum n a_n X^{n-1}$. Give an example where the regions of convergence are not the same.

15. (a) Find an example of an infinite sum of nonzero rational numbers which converges in $|\ |_p$ for *every* p and which converges in the reals (i.e., in $|\ |_\infty = |\ |$).
(b) Can such a sum ever converge to a rational number in any $|\ |_p$ or $|\ |_\infty$?

16. Suppose that, instead of dealing with power series, we decided to mimic the familiar definition of differentiable functions and say that a function $f: \Omega \to \Omega$ is "differentiable" at $a \in \Omega$ if $(f(x) - f(a))/(x - a)$ approaches a limit in Ω as $|x - a|_p \to 0$. First of all, prove that, if $f(X) = \sum_{n=0}^{\infty} a_n X^n$ is a power series, then it is differentiable at every point in its disc of convergence, and it can be differentiated term-by-term, i.e., its derivative at a point a in the disc of convergence is equal to $\sum_{n=1}^{\infty} n a_n a^{n-1}$. In other words, the derivative function is the formal derivative power series.

17. Using the definition of "differentiable" in the previous problem, give an example of a function $f: \Omega \to \Omega$ which is everywhere differentiable, has derivative identically zero, but is not locally constant (see discussion of locally constant functions at the beginning of §II.3). This example can be made to vanish along with all of its derivatives at $x = 0$, but not be constant in any neighborhood of 0. Thus, it is in the spirit of the wonderful function e^{-1/x^2} from real calculus, which does not equal its (identically zero) Taylor series at the origin.

18. The Mean Value Theorem of ordinary calculus, applied to $f: \mathbb{R} \to \mathbb{R}$, $f(x) = x^p - x$ on the interval $\{x \in \mathbb{R} \mid |x| \le 1\}$, says that, since $f(1) = f(-1) = 0$ (here we are assuming that $p > 2$), we must have

$$f'(\alpha) = 0 \quad \text{for some} \quad \alpha \in \mathbb{R}, |\alpha| \le 1.$$

(In fact, $\alpha = \pm(1/p)^{1/(p-1)}$ works.) Does this hold with $\mathbb{R}$ replaced by Ω and $|\ |$ replaced by $|\ |_p$?

19. Let $f: \mathbb{Q}_p \to \mathbb{Q}_p$ be defined by $x = \sum a_n p^n \mapsto \sum g(a_n)p^n$, where $\sum a_n p^n$ is the p-adic expansion of x and $g: \{0, 1, \ldots, p-1\} \to \mathbb{Q}_p$ is any function. Prove that f is continuous. If $g(a) = a^2$ and $p \ne 2$, prove that f is *nowhere* differentiable.

20. Prove that for any N and for any $j = 1, 2, \ldots, N$,

$$(1 + X)^{p^N} - 1 \in p^j \mathbb{Z}[X] + X^{p^{N-j+1}} \mathbb{Z}[X].$$

Suppose that a/b is a rational number with $|a/b|_p \le 1$, and you want to find the first M coefficients (M is a large number) of the power series $(1 + X)^{a/b}$ to a certain p-adic accuracy. Discuss how to write a simple algorithm (e.g., a computer program) to do this. (Only do arithmetic in $\mathbb{Z}/p^n\mathbb{Z}$, not in $\mathbb{Q}$, since the former is generally much easier to do by computer.)

21. If R is any ring, define the ring $R[[X_1, \ldots, X_n]]$ (abbreviated $R[[X]]$) of formal power series in n variables as the set of all sequences $\{r_{i_1, \cdots, i_n}\}$ indexed by n-tuples $i_1, \ldots, i_n$ of nonnegative integers (such a sequence is thought of as $\sum r_{i_1, \cdots, i_n} X_1^{i_1} \cdots X_n^{i_n}$ and sometimes abbreviated $\sum r_i X^i$), with addition and multiplication defined in the usual way. Thus, $\{r_{i_1, \cdots, i_n}\} + \{s_{i_1, \cdots, i_n}\} = \{t_{i_1, \cdots, i_n}\}$, where $t_{i_1, \cdots, i_n} = r_{i_1, \cdots, i_n} + s_{i_1, \cdots, i_n}$; and $\{r_{i_1, \cdots, i_n}\} \cdot \{s_{i_1, \cdots, i_n}\} = \{t_{i_1, \cdots, i_n}\}$, where $t_{i_1, \cdots, i_n} = \sum r_{j_1, \cdots, j_n} s_{k_1, \cdots, k_n}$ with the summation taken over all pairs of n-tuples $j_1, \ldots, j_n$ and $k_1, \ldots, k_n$ for which $j_1 + k_1 = i_1$, $j_2 + k_2 = i_2, \ldots, j_n + k_n = i_n$.

By the minimal total degree deg f of a nonzero power series f we mean the least d such that some $r_{i_1,\cdots,i_n}$ with $i_1 + i_2 + \cdots + i_n = d$ is nonzero. We can define a topology, the "X-adic topology," on $R[[X]]$ by fixing some positive real number $\rho < 1$ and defining the "X-adic norm" by

$$|f|_X \underset{\text{def}}{=} \rho^{\deg f} \qquad (|0|_X \text{ is defined to be } 0).$$

(1) Show that $|\ |_X$ makes $R[[X]]$ into a non-Archimedean metric space (see the first definition in §1.1; by "non-Archimedean," we mean, of course, that the third condition can be replaced by: $d(x, y) \leq \max(d(x, z), d(z, y))$). Say in words what it means for $|f|_X$ to be < 1.

(2) Show that $R[[X]]$ is complete with respect to $|\ |_X$.

(3) Show that an infinite product of series $f_j \in R[[X]]$ converges to a nonzero number if and only if none of the f_j is zero and $|f_j - 1|_X \to 0$ (where 1 is the constant power series $\{r_{i_1,\ldots,i_n}\}$ for which $r_{0,\ldots,0} = 1$ and all other $r_{i_1,\ldots,i_n} = 0$). We will use this in §2 to see that the horrible power series defined at the end of that section makes sense.

(4) If $f \in R[[X]]$, define f_d to be the same as f but with all coefficients $r_{i_1,\cdots,i_n}$ with $i_1 + \cdots + i_n > d$ replaced by 0. Thus, f_d is a polynomial in n variables. Let $g_1, \ldots, g_n \in R[[X]]$. Note that $f_d(g_1(X), g_2(X), \ldots, g_n(X))$ is well-defined for every d, since it's just a finite sum of products of power series. Prove that $\{f_d(g_1(X), \ldots, g_n(X))\}_{d=0,1,2,\ldots}$ is a Cauchy sequence in $R[[X]]$ if $|g_j|_X < 1$ for $j = 1, \ldots, n$. In that case call its limit $f \circ g$.

(5) Now let R be the field $\mathbb{R}$ of real numbers, and suppose that $f, f_d, g_1, \ldots,$ g_n are as in (4), with $|g_j|_X < 1$. Further suppose that for some $\varepsilon > 0$ the series f and all of the series g_j are absolutely convergent whenever we substitute $X_i = x_i$ in the interval $[-\varepsilon, \varepsilon] \subset \mathbb{R}$. Prove that the series $f \circ g$ is absolutely convergent whenever we substitute $X_i = x_i$ in the (perhaps smaller) interval $[-\varepsilon', \varepsilon']$ for some $\varepsilon' > 0$.

(6) Under the conditions in (5), prove that if $f \circ g(x_1, \ldots, x_n)$ has value 0 for every choice of $x_1, \ldots, x_n \in [-\varepsilon', \varepsilon']$, then $f \circ g$ is the zero power series in $\mathbb{R}[[X]]$.

(7) As an example, let $n = 3$, write X, Y, Z instead of X_1, X_2, X_3, and let

$$f(X, Y, Z) = \sum_{i=1}^{\infty} (-1)^{i+1}(X^i/i + Y^i/i - Z^i/i),$$

$$g_1(X, Y, Z) = X,$$
$$g_2(X, Y, Z) = Y,$$
$$g_3(X, Y, Z) = X + Y + XY.$$

As another example, let $n = 2$,

$$f(X, Y) = \left(\sum_{i=1}^{\infty} (-1)^{i+1} X^i/i\right) - Y,$$

$$g_1(X, Y) = \sum_{i=1}^{\infty} X^i/i!,$$

$$g_2(X, Y) = X.$$

Explain how your result in (6) can be used to prove the basic facts about the elementary p-adic power series. (Construct the f and g_j for one or two more cases.)

2. The logarithm, gamma and Artin–Hasse exponential functions

In this section we look at some further examples of p-adic analytic functions (more precisely, "locally" analytic functions) which have proven useful in studying various questions in number theory. The first is Iwasawa's extension of the logarithm.

Recall that the Taylor series $\log_p x = \sum_{n=1}^{\infty} (-1)^{n+1}(x-1)^n/n$ converges in the open unit disc around 1. The following proposition says that there is a unique function extending $\log_p x$ to all nonzero x and having certain convenient properties.

Proposition. *There exists a unique function* $\log_p : \Omega^{\times} \to \Omega$ *(where* $\Omega^{\times} = \Omega - \{0\}$*) such that:*

(1) $\log_p x$ *agrees with the earlier definition for* $|x - 1|_p < 1$, *i.e.,*

$$\log_p x = \sum_{n=1}^{\infty} (-1)^{n+1}(x-1)^n/n \quad for \ |x - 1|_p < 1;$$

(2) $\log_p (xy) = \log_p x + \log_p y$ *for all* $x, y \in \Omega^{\times}$;

(3) $\log_p p = 0$.

PROOF. Recall from §III.4 that any nonzero $x \in \Omega$ can be written in the form $x = p^r \omega(x_1)\langle x_1 \rangle$, where p^r is some fixed root of the equation $x^b - p^a = 0$, with $r = a/b = \mathrm{ord}_p x$, $\omega(x_1)$ is a root of unity, and $|\langle x_1 \rangle - 1|_p < 1$. There is thus only one possible way to define $\log_p x$ consistently with (1)–(3). Namely, (2) and (3) imply that $\log_p(p^r) = \log_p(\omega(x_1)) = 0$, and hence we must have

$$\log_p x = \sum_{n=1}^{\infty} (-1)^{n+1}(\langle x_1 \rangle - 1)^n/n.$$

We thus know that there is *at most* one definition of $\log_p x$ which has the desired properties, namely, the definition $\log_p x = \log_p \langle x_1 \rangle$. It remains to show that the three desired properties are actually satisfied. Properties (1) and (3) are obvious from the definition.

In the course of our definition of $\log_p x$, we made a rather arbitrary choice of a bth root of p^a. But if we had chosen another bth root of p^a for our p^r, this would have altered x_1 by a bth root of unity and hence would have altered $\omega(x_1)$ and $\langle x_1 \rangle$ by certain roots of unity. Notice that the new $\langle x_1' \rangle$ would have to differ from the old $\langle x_1 \rangle$ by a pth power root of unity, because $\zeta = \langle x_1' \rangle / \langle x_1 \rangle$ is in the open unit disc about 1 (see Exercise 7 in §III.4). In any case, the definition $\log_p x = \log_p \langle x_1 \rangle$ would not be affected by this replacement of x_1 by x_1', because $\log_p \zeta = 0$, as remarked in §1. Thus, our definition really does not depend on the choice of p^r.

We now prove property (2). Let $x = p^r \omega(x_1)\langle x_1 \rangle$, $y = p^s \omega(y_1)\langle y_1 \rangle$, $z = xy = p^{r+s}\omega(z_1)\langle z_1 \rangle$. Now p^{r+s} is not necessarily the same fractional power of p as $p^r p^s$; it may differ by a root of unity. But the definition of $\log_p z$ does not change if we change our choice of p^{r+s} to $p^r p^s$. In that case, $z_1 = z/p^r p^s = x_1 y_1$, and so $\langle z_1 \rangle = \langle x_1 \rangle \langle y_1 \rangle$, and

$$\log_p z = \log_p \langle z_1 \rangle = \log_p \langle x_1 \rangle + \log_p \langle y_1 \rangle = \log_p x + \log_p y,$$

where the middle equality was proved in the last section in our discussion of the power series $\sum (-1)^{n+1} x^n / n$. This completes the proof of the proposition.
$\square$

Now let $x_0 \neq 0$ be a fixed point of Ω. Let $r = |x_0|_p$, and suppose that x is in the largest disc about x_0 which does not contain zero, i.e., $D_{x_0}(r^-)$. Then $|x/x_0 - 1|_p < 1$, and so

$$\log_p x = \log_p(x_0(1 + x/x_0 - 1)) = \log_p x_0 + \sum_{n=1}^{\infty} (-1)^{n+1}(x - x_0)^n / n x_0^n.$$

Thus, in $D_{x_0}(r^-)$ the function $\log_p x$ can be represented by a convergent power series in $x - x_0$. Whenever a function can be represented by a convergent power series in a neighborhood of any point in its region of definition, we say that it is *locally analytic*. Thus, $\log_p x$ is a locally analytic function on $\Omega - \{0\}$.

Recall from Exericse 16 of §1 that the usual definition of the derivative can be applied to *p*-adic functions, and that power series are always differentiable in their region of convergence, with the derivative obtained by term-by-term differentiation. In particular, applying this to $\log_p x$ in $D_{x_0}(r^-)$, we obtain

$$\frac{d}{dx} \log_p x = \sum_{n=1}^{\infty} (-1)^{n+1}(x - x_0)^{n-1} / x_0^n$$

$$= x_0^{-1} \sum_{n=0}^{\infty} (1 - x/x_0)^n$$

$$= x_0^{-1} / (x/x_0) = 1/x$$

for $x \in D_{x_0}(r^-)$. We conclude:

Proposition. $\log_p x$ *is locally analytic on* $\Omega - \{0\}$ *with derivative* $1/x$.

The next function we discuss is the *p*-adic analogue of the gamma-function.

The classical gamma-function is a function from $\mathbb{R}$ to $\mathbb{R}$ which "interpolates" $n!$ (actually, $\Gamma(s)$ is defined for complex s, but we aren't interested in that here). More precisely, it is a continuous function of a real variable s excluding $s = 0, -1, -2, -3, \ldots$ (where it has "poles") which satisfies

$$\Gamma(s + 1) = s! \quad \text{for} \quad s = 0, 1, 2, 3, \ldots.$$

Since the positive integers are not dense in $\mathbb{R}$, there are infinitely many functions which satisfy this equality; but there is only one which has certain other convenient properties. This gamma-function can be defined for $s > 0$ by:

$$\Gamma(s) = \int_0^\infty x^{s-1} e^{-x}\, dx.$$

Thus, the gamma-function is the "Mellin transform" of e^{-x} (see §7 of Chapter II). It is not hard to check (see Exercises 6–7 below) that this improper integral converges for $s > 0$, and that the function $\Gamma(s)$ defined in this way satisfies $\Gamma(s+1) = s\Gamma(s)$ for all $s > 0$. In addition, $\Gamma(1) = \int_0^\infty e^{-x}\, dx = 1$; then $\Gamma(s+1) = s\Gamma(s) = s(s-1)\Gamma(s-1) = \cdots = s!\Gamma(1) = s!$, so this function really is an interpolation of the factorial function.

We would now like to do something similar p-adically, i.e., find a continuous function from $\mathbb{Z}_p$ to $\mathbb{Z}_p$ whose values at positive integers $s+1$ coincide with $s!$.

We shall assume that $p > 2$ in what follows; minor modifications are needed if $p = 2$.

Recall from §2 of Chapter II under what conditions a function $f(s)$ on the positive integers can be interpolated to all of $\mathbb{Z}_p$. Such a continuous interpolation exists if and only if for every $\varepsilon > 0$ there exists N such that

$$s \equiv s' \pmod{p^N} \quad \text{implies} \quad |f(s) - f(s')|_p < \varepsilon. \tag{*}$$

In that case the interpolating function is unique and is defined by

$$f(s) = \lim_{k \to s,\, k \in \mathbb{N}} f(k).$$

Unfortunately, the basic condition (*) does not hold for $f(s) = (s-1)!$, since, for example, $|f(1) - f(1 + p^N)|_p = 1$ for any $N > 0$, since p divides $s!$ whenever $s \geq p$. The problem is that, whenever s is a large integer in the old-fashioned archimedean sense, $s!$ is divisible by a large power of p, i.e., $f(s) \to 0$ p-adically as $s \to \infty$.

We could modify the factorial function in a way analogous to how we modified the Riemann zeta-function in Chapter II ("removing the Euler factor") by discarding indices divisible by p. That is, we could try to interpolate the $f(s)$ defined by:

$$f(s+1) = \prod_{j \leq s,\, p \nmid j} j = \frac{s!}{[s/p]!\, p^{[s/p]}}.$$

However, once again we have problems (see Exercise 8 below). But if we modify $f(s)$ one final time by a mere change in sign for odd s, we can then interpolate.

Proposition. *Let*

$$\Gamma_p(k) = (-1)^k \prod_{j < k,\, p \nmid j} j, \qquad k = 1, 2, 3, \ldots.$$

(*When $k = 1$, the empty product is defined to be 1, i.e., $\Gamma_p(1) = -1$.*)

Then Γ_p extends uniquely to a continuous function $\Gamma_p \colon \mathbb{Z}_p \to \mathbb{Z}_p^\times$ defined by

$$\Gamma_p(s) = \lim_{k \to s, \, k \in \mathbb{N}} (-1)^k \prod_{j < k, \, p \nmid j} j.$$

PROOF. It suffices to prove (∗); in fact we shall prove that

$$k' = k + k_1 p^N \quad \text{implies} \quad \Gamma_p(k) \equiv \Gamma_p(k') \pmod{p^N}.$$

Notice that $\Gamma_p(k) \in \mathbb{Z}_p^\times$ (that is why Γ_p will be a map from $\mathbb{Z}_p$ to $\mathbb{Z}_p^\times$ as soon as we show that the continuous interpolation exists). Hence the right side of the above implication is equivalent to the congruence

$$1 \equiv \Gamma_p(k')/\Gamma_p(k) = (-1)^{k'-k} \prod_{k \le j < k', \, p \nmid j} j \pmod{p^N}.$$

If we prove this for $k_1 = 1$, i.e., for $k' = k + p^N$, then by multiplying together the congruences with k replaced by $k + ip^N$ ($i = 0, \ldots, k_1 - 1$)

$$1 \equiv (-1)^{k+(i+1)p^N-(k+ip^N)} \prod_{k+ip^N \le j < k+(i+1)p^N, \, p \nmid j} j \pmod{p^N},$$

we immediately obtain the desired congruence. Since p is odd, we have $(-1)^{p^N} = -1$, and so we have reduced the proof to showing that

$$\prod_{k \le j < k+p^N, \, p \nmid j} j \equiv -1 \pmod{p^N}.$$

Since the product runs through every congruence class in $(\mathbb{Z}/p^N\mathbb{Z})^\times$ exactly once, we have

$$\prod_{k \le j < k+p^N, \, p \nmid j} j \equiv \prod_{0 < j < p^N, \, p \nmid j} j \pmod{p^N}.$$

Thus, it remains to prove that the product on the right is $\equiv -1 \pmod{p^N}$. We now pair off elements j and j' which satisfy $jj' \equiv 1 \pmod{p^N}$. For each j there is precisely one such j'. Since $p > 2$, there are only two values of j for which $j' = j$, i.e., for which $j^2 \equiv 1 \pmod{p^N}$ (see Exercise 9 below). Thus,

$$\prod_{0 < j < p^N, \, p \nmid j} \equiv (\textstyle\prod jj')(1)(-1) \equiv -1 \pmod{p^N},$$

as desired. □

The key step in the proof, the congruence for $\prod_{j < p^N, \, p \nmid j} j$, is a generalization of Wilson's theorem, which is the case $N = 1 \colon (p-1)! \equiv -1 \pmod{p}$.

Basic properties of Γ_p.

$$\frac{\Gamma_p(s+1)}{\Gamma_p(s)} = \begin{cases} -s & \text{if } s \in \mathbb{Z}_p^\times; \\ -1 & \text{if } s \in p\mathbb{Z}_p. \end{cases} \tag{1}$$

PROOF. Since both sides are continuous functions from $\mathbb{Z}_p$ to $\mathbb{Z}_p^\times$, it suffices to check equality on the dense subset $\mathbb{N}$, i.e., when $s = k \in \mathbb{N}$. But then it follows immediately from the definition of $\Gamma_p(k)$. □

(2) If $s \in \mathbb{Z}_p$, write $s = s_0 + ps_1$, where $s_0 \in \{1, 2, \ldots, p\}$ is the first digit in s unless $s \in p\mathbb{Z}_p$, in which case $s_0 = p$ rather than 0. Then

$$\Gamma_p(s)\Gamma_p(1 - s) = (-1)^{s_0}.$$

PROOF. Again by continuity it suffices to check this when $s = k$. For $s = 1$ the equality holds because $\Gamma_p(1) = -1$ by definition, and $\Gamma_p(0) = -\Gamma_p(1) = 1$ by property (1). Now use induction, assuming the equality for $s = k$ and then proving it for $k + 1$. Using property (1), we have

$$\frac{\Gamma_p(s + 1)\Gamma_p(1 - (s + 1))}{\Gamma_p(s)\Gamma_p(1 - s)} = \begin{cases} -s/(-(-s)) = -1 & \text{if } s \in \mathbb{Z}_p^{\times}; \\ -1/(-1) = 1 & \text{if } s \in p\mathbb{Z}_p, \end{cases}$$

and this shows that the equality in (2) for $s + 1$ follows from the equality for s. $\qquad\square$

(3) For $s \in \mathbb{Z}_p$, define s_0 and s_1 as in property (2). Let m be any positive integer not divisible by p. Then

$$\frac{\prod_{h=0}^{m-1} \Gamma_p((s + h)/m)}{\Gamma_p(s) \prod_{h=1}^{m-1} \Gamma_p(h/m)} = m^{1 - s_0}(m^{-(p-1)})^{s_1}.$$

Remarks. 1. The expression on the right makes sense, because the number being raised to the p-adic power, namely $m^{-(p-1)}$, is congruent to 1 mod p. (See §2 of Chapter II.) Of course, s_0 is a positive integer, so m^{1-s_0} makes sense.

2. The classical gamma-function can be shown to satisfy the "Gauss–Legendre multiplication formula"

$$\frac{\prod_{h=0}^{m-1} \Gamma((s + h)/m)}{\Gamma(s) \prod_{h=1}^{m-1} \Gamma(h/m)} = m^{1-s}.$$

PROOF OF (3). Let $f(s)$ be the left side and let $g(s)$ be the right side of the equation. Both f and g are continuous, so it suffices to check equality for $s = k \in \mathbb{N}$. For $s = k = 1$ both sides are clearly 1. We proceed by induction on k. We have

$$\frac{f(s + 1)}{f(s)} = \frac{\Gamma_p(s)\Gamma_p((s/m) + 1)}{\Gamma_p(s + 1)\Gamma_p(s/m)} = \begin{cases} 1/m & \text{if } s \in \mathbb{Z}_p^{\times}; \\ 1 & \text{if } s \in p\mathbb{Z}_p. \end{cases}$$

On the other hand, if $s \in \mathbb{Z}_p^{\times}$, we have $g(s + 1)/g(s) = 1/m$, since then $(s + 1)_0 = s_0 + 1$ and $(s + 1)_1 = s_1$, while if $s \in p\mathbb{Z}_p$ we have $g(s + 1)/g(s) = 1$, since then $(s + 1)_0 = s_0 - (p - 1)$ and $(s + 1)_1 = s_1 + 1$. Hence $f(s + 1)/f(s) = g(s + 1)/g(s)$, and the induction step follows. $\qquad\square$

This concludes our discussion of the p-adic gamma-function.

We now introduce an "elementary function" which is "better" than $\exp_p$—has a larger disc of convergence—and which can often be used to play a similar role to exp in situations when better convergence than $D(p^{-1/(p-1)}-)$ is needed. To do this, we first give an infinite product formula for the ordinary exponential function, in terms of the "Möbius function" $\mu(n)$, which is often used in number theory. For $n \in \{1, 2, 3, \ldots\}$ we define

$$\mu(n) = \begin{cases} 0, & \text{if } n \text{ is divisible by a perfect square greater than } 1; \\ (-1)^k, & \text{if } n \text{ is a product of } k \text{ distinct prime factors.} \end{cases}$$

Thus, $1 = \mu(1) = \mu(6) = \mu(221) = \mu(1155), 0 = \mu(9) = \mu(98), -1 = \mu(2) = \mu(97) = \mu(30) = \mu(105)$. A basic fact about μ is that the sum of the values of μ over the divisors of a positive integer n equals 1 if $n = 1$ and 0 otherwise. This is true because, if $n = p_1^{a_1} \cdots p_s^{a_s}$ is the decomposition into prime factors, and if $s \geq 1$, then we have:

$$\sum_{d|n} \mu(d) = \sum_{\substack{\text{all possible} \\ \varepsilon_i = 0 \text{ or } 1, i = 1, \cdots, s}} \mu(p_1^{\varepsilon_1} \cdots p_s^{\varepsilon_s}) = \sum (-1)^{\sum \varepsilon_i} = (1 - 1)^s = 0.$$

We now claim that the following "formal identity" holds in $\mathbb{Q}[[X]]$:

$$\exp(X) = \prod_{n=1}^{\infty} (1 - X^n)^{-\mu(n)/n} \underset{\text{def}}{=} \prod_{n=1}^{\infty} B_{-\mu(n)/n, p}(-X^n).$$

(Note that this infinite product of infinite series makes sense, since the nth series starts with $1 + \mu(n)/nX^n$, i.e., has no powers of X less than the nth; thus, only finitely many series have to be multiplied together to determine the coefficient of any given power of X.) To prove this, take the log of the right hand side. You get:

$$\log \prod_{n=1}^{\infty} (1 - X^n)^{-\mu(n)/n} = -\sum_{n=1}^{\infty} \frac{\mu(n)}{n} \log(1 - X^n) = \sum_{n=1}^{\infty} \frac{\mu(n)}{n} \sum_{m=1}^{\infty} \frac{X^{nm}}{m}$$

$$= \sum_{j=1}^{\infty} \left[\frac{X^j}{j} \sum_{n|j} \mu(n) \right] \qquad (j = \text{old } mn),$$

gathering together coefficients of the same power of X. By the basic property of μ proved above, this equals X. Taking exp of both sides, we obtain the desired formal identity.

(Several times we have used the principle, mentioned in the discussion of $\log_p$ and developed in Exercise 21 of the last section, that manipulation of formal power series as though the variables were real numbers is justified as long as the series involved all converge in some interval about 0.)

If we look at $\prod_{n=1}^{\infty} (1 - X^n)^{-\mu(n)/n}$ p-adically, we can focus in on where the "trouble" comes in. By "trouble" I mean why it only converges on $D(p^{-1/(p-1)}-)$ and not on $D(1^-)$. Namely, if $p|n$ and n is square-free, then $(1 - X^n)^{-\mu(n)/n}$ only converges when an x is substituted for which

$$|x^n|_p = |x|_p^n \in D(r^-), \quad \text{where } r = p^{-1/(p-1)} \bigg/ \left| -\frac{\mu(n)}{n} \right|_p = p^{-1/(p-1)} |n|_p.$$

For example, if $n = p$, then this converges precisely when

$$|x|_p < \left(p^{-1/(p-1)}\frac{1}{p}\right)^{1/p} = p^{-1/(p-1)}.$$

But as long as $p \nmid n$ we're O.K.: that is, since $-\mu(n)/n \in \mathbb{Z}_p$, we have $(1 - X^n)^{-\mu(n)/n} \in \mathbb{Z}_p[[X]]$. (Remember in all this that $(1 - X^n)^a$ is just shorthand for $B_{a,p}(-X^n) = \sum_{i=0}^{\infty} a(a-1)\cdots(a-i+1)/i!\,(-X^n)^i$.)

So let's define a new function E_p, which we call the "Artin–Hasse exponential," by just forgetting about the "bad" terms in the infinite product (this is very similar to our "removing the Euler factor" in order to define the p-adic zeta-function in Chapter II):

$$E_p(X) \underset{\text{def}}{=} \prod_{\substack{n=1 \\ p \nmid n}}^{\infty} (1 - X^n)^{-\mu(n)/n} \in \mathbb{Q}[[X]].$$

Since each infinite series $B_{-\mu(n)/n,p}(-X^n)$ is in $1 + X^n \mathbb{Z}_p[[X]]$, their infinite product makes sense (only finitely many have to be multiplied to get the coefficient of any given power of X), and it lies in $1 + X\mathbb{Z}_p[[X]]$.

We can easily find a simpler expression for $E_p(X)$, using the property of the μ function:

$$\sum_{d|n,\,p\nmid d} \mu(d) = \begin{cases} 1 & \text{if } n \text{ is a power of } p; \\ 0 & \text{otherwise.} \end{cases}$$

This property follows immediately from the earlier property of μ, applied to $n/p^{\mathrm{ord}_p n}$ in place of n. Considering $E_p(X)$ over $\mathbb{R}$ (or $\mathbb{C}$) and taking the logarithm as before gives:

$$\log E_p(X) = -\sum_{\substack{n=1 \\ p\nmid n}}^{\infty} \frac{\mu(n)}{n} \sum_{m=1}^{\infty} \frac{X^{mn}}{m} = \sum_{j=1}^{\infty} \left[\frac{X^j}{j} \sum_{n|j,\,p\nmid n} \mu(n)\right]$$

$$= \sum_{m=0}^{\infty} X^{p^m}/p^m.$$

Hence,

$$E_p(X) = \exp\left(X + \frac{X^p}{p} + \frac{X^{p^2}}{p^2} + \frac{X^{p^3}}{p^3} + \cdots\right),$$

as an equality of formal power series in $\mathbb{Q}[[X]]$.

The important thing about $E_p(X)$, in distinction from $\exp_p(X)$, is that $E_p(X) \in \mathbb{Z}_p[[X]]$. Thus, $E_p(X)$ converges in $D(1^-)$. It can be seen (Exercise 11 of §IV.4) that this is its *exact* disc of convergence, i.e., it does not converge on $D(1)$.

We conclude this section with a useful general lemma, due to Dwork.

Lemma 3. *Let* $F(X) = \sum a_i X^i \in 1 + X\mathbb{Q}_p[[X]]$. *Then* $F(X) \in 1 + X\mathbb{Z}_p[[X]]$ *if and only if* $F(X^p)/(F(X))^p \in 1 + pX\mathbb{Z}_p[[X]]$.

PROOF. If $F(X) \in 1 + X\mathbb{Z}_p[[X]]$, then, since $(a+b)^p \equiv a^p + b^p \pmod{p}$ and $a^p \equiv a \pmod{p}$ for $a \in \mathbb{Z}_p$, it follows that

$$(F(X))^p = F(X^p) + pG(X) \quad \text{for some } G(X) \in X\mathbb{Z}_p[[X]].$$

Hence

$$\frac{F(X^p)}{(F(X))^p} = 1 - \frac{pG(X)}{(F(X))^p} \in 1 + pX\mathbb{Z}_p[[X]],$$

because $(F(X))^p \in 1 + X\mathbb{Z}_p[[X]]$ and hence can be inverted (see Exercise 3 of §1).

In the other direction, write

$$F(X^p) = (F(X))^p G(X), \qquad G(X) \in 1 + pX\mathbb{Z}_p[[X]],$$

$$G(X) = \sum b_i X^i, \qquad F(X) = \sum a_i X^i.$$

We prove by induction that $a_i \in \mathbb{Z}_p$. By assumption, $a_0 = 1$. Suppose $a_i \in \mathbb{Z}_p$ for $i < n$. Then, equating coefficients of X^n on both sides gives

$$\left. \begin{matrix} a_{n/p} & \text{if } p \text{ divides } n \\ 0 & \text{otherwise} \end{matrix} \right\} = \text{coefficient of } X^n \text{ in } \left(\sum_{i=0}^{n} a_i X^i \right)^p \left(1 + \sum_{i=1}^{n} b_i X^i \right).$$

If we expand the polynomial on the right, subtract $a_{n/p}$ in the case $p|n$ (and recall that $a_{n/p} \equiv a_{n/p}^p \bmod p$), and notice that the resulting expression consists of pa_n added to a bunch of terms in $p\mathbb{Z}_p$, we can conclude that $pa_n \in p\mathbb{Z}_p$, i.e., $a_n \in \mathbb{Z}_p$. $\square$

Dwork's lemma can be used to give an easy direct proof (without using the infinite product expansion) that the formal power series $E_p(X) = e^{X + (X^p/p) + (X^{p^2}/p^2) + \cdots}$ has coefficients in $\mathbb{Z}_p$ (see Exercise 17 below).

Dwork's lemma, which seems a little bizarre at first glance, is actually an example of a deep phenomenon in *p*-adic analysis. It says that if we know something about $F(X^p)/(F(X))^p$, then we know something about F. This quotient expression $F(X^p)/(F(X))^p$ measures how much difference there is between raising X to the pth power and then applying F, versus applying F and then raising to the pth power, i.e., it measures how far off F is from commuting with the pth power map. The pth power map plays a crucial role, as we've seen in other *p*-adic contexts (recall the section on finite fields). So Dwork's lemma says that if F "commutes to within mod p" with the pth power map, i.e., $F(X^p)/(F(X))^p = 1 + p \cdot \sum (p\text{-adic integers}) X^i$, then F has *p*-adic integer coefficients.

We apply this lemma to a function that will come up in Dwork's proof of the rationality of the zeta-function. First, note that Lemma 3 can be generalized as follows: Let $F(X, Y) = \sum a_{m,n} X^n Y^m$ be a power series in *two* variables X and Y with constant term 1, i.e.,

$$F(X, Y) \in 1 + X\mathbb{Q}_p[[X, Y]] + Y\mathbb{Q}_p[[X, Y]].$$

Then all the $a_{m,n}$'s are in $\mathbb{Z}_p$ if and only if

$$F(X^p, Y^p)/(F(X, Y))^p \in 1 + pX\mathbb{Z}_p[[X, Y]] + pY\mathbb{Z}_p[[X, Y]].$$

The proof is completely analogous to the proof of Lemma 3.

We now define a series $F(X, Y)$ in $\mathbb{Q}[[X, Y]]$ as follows:

$$F(X, Y) = B_{X,p}(Y)B_{(X^p - X)/p,p}(Y^p)B_{(X^{p^2} - X^p)/p^2,p}(Y^{p^2})\cdots B_{(X^{p^n} - X^{p^{n-1}})/p^n,p}(Y^{p^n})\cdots$$

$$= (1 + Y)^X(1 + Y^p)^{(X^p - X)/p}(1 + Y^{p^2})^{(X^{p^2} - X^p)/p^2}\cdots$$

$$\times (1 + Y^{p^n})^{(X^{p^n} - X^{p^{n-1}})/p^n}\cdots$$

$$= \sum_{i=0}^{\infty} \frac{X(X - 1)\cdots(X - i + 1)}{i!} Y^i$$

$$\prod_{n=1}^{\infty}\left(\sum_{i=0}^{\infty} \frac{X^{p^n} - X^{p^{n-1}}}{p^n}\left(\frac{X^{p^n} - X^{p^{n-1}}}{p^n} - 1\right)\cdots\right.$$

$$\left.\times\left(\frac{X^{p^n} - X^{p^{n-1}}}{p^n} - i + 1\right)\frac{Y^{ip^n}}{i!}\right).$$

Since we only have to take finitely many terms in the product to get the co-efficient of any $X^n Y^m$, this is a well-defined infinite series $F(X, Y) = \sum a_{m,n}X^n Y^m$ in $1 + X\mathbb{Q}_p[[X, Y]] + Y\mathbb{Q}_p[[X, Y]]$. We use the generalization of Lemma 3 to prove that $a_{m,n} \in \mathbb{Z}_p$. Namely, we have

$$\frac{F(X^p, Y^p)}{(F(X, Y))^p} = \frac{(1 + Y^p)^{X^p}(1 + Y^{p^2})^{(X^{p^2} - X^p)/p}(1 + Y^{p^3})^{(X^{p^3} - X^{p^2})/p^2}\cdots}{(1 + Y)^{pX}(1 + Y^p)^{X^p - X}(1 + Y^{p^2})^{(X^{p^2} - X^p)/p}\cdots}$$

$$= \frac{(1 + Y^p)^X}{(1 + Y)^{pX}}.$$

We must show that $(1 + Y^p)^X/(1 + Y)^{pX}$ is in $1 + pX\mathbb{Z}_p[[X, Y]] + pY\mathbb{Z}_p[[X, Y]]$. Applying Lemma 3 in the other direction shows that, since $1 + Y \in 1 + Y\mathbb{Z}_p[[Y]]$, it follows that

$$(1 + Y^p)/(1 + Y)^p = 1 + pYG(Y), G(Y) \in \mathbb{Z}_p[[Y]].$$

Thus,

$$\frac{(1 + Y^p)^X}{(1 + Y)^{pX}} = (1 + pYG(Y))^X = \sum_{i=0}^{\infty} \frac{X(X - 1)\cdots(X - i + 1)}{i!} p^i(YG(Y))^i,$$

which is clearly in $1 + pX\mathbb{Z}_p[[X, Y]] + pY\mathbb{Z}_p[[X. Y]]$. We conclude that $F(X, Y) \in \mathbb{Z}_p[[X, Y]]$.

EXERCISES

1. Find $\log_7 42 \bmod 7^4$ and $\log_2 15 \bmod 2^{12}$.

2. Prove that the image of $\mathbb{Z}_p$ under $\log_p$ is $p\mathbb{Z}_p$ for $p > 2$ and is $4\mathbb{Z}_2$ for $p = 2$.

3. For $p > 2$ and $a \in \mathbb{Z}_p^{\times}$, prove that p^2 divides $\log_p a$ if and only if $a^{p-1} \equiv 1 \bmod p^2$.

4. Find the derivative of the locally analytic function $x \log_p x - x$.

5. Suppose that a function $f: \Omega^{\times} \to \Omega$ satisfies properties (1) and (2) of the proposition at the beginning of this section. Prove that $f(x)$ must be of the form $f(x) = \log_p x + c\, \mathrm{ord}_p x$ for some constant $c \in \Omega$.

6. Verify that the improper integral $\int_0^\infty x^{s-1}e^{-x}\,dx$ converges if and only if $s > 0$.

7. Using integration by parts, prove that $\Gamma(s + 1) = s\Gamma(s)$, where $\Gamma(s)$ is defined by the integral in Exercise 6.

8. Show that the function $f(s) = \prod_{j<s,\,p\nmid j} j$ for $s \in \mathbb{N}$ does *not* extend to a continuous function on $\mathbb{Z}_p$.

9. For $p > 2$, show that if $j^2 - 1 \equiv 0 \bmod p^N$, then $j \equiv \pm 1 \bmod p^N$. What happens if $p = 2$?

10. Show that $\Gamma_p(1/2)^2 = -(\frac{-1}{p})$, where $(\frac{-1}{p}) = 1$ if $x^2 = -1$ has a solution in $\mathbb{F}_p$ and it equals -1 otherwise.

11. Compute $\Gamma_5(1/4)$ and $\Gamma_7(1/3)$ to four digits (if you don't have a computer or programmable calculator handy, then compute them to two digits).

12. Let $\sqrt{-1} \in \mathbb{Z}_5$ denote the root with first digit 3, and let $\sqrt{-3} \in \mathbb{Z}_7$ be the root with first digit 2. Use Exercise 9 of §I.5 and Exercise 11 above to verify the following equalities to 4 digits:

$$\Gamma_5(1/4)^2 = -2 + \sqrt{-1}; \; \Gamma_7(1/3)^3 = (1 - 3\sqrt{-3})/2.$$

Note: These equalities are known to be true, but no down-to-earth proof (without *p*-adic cohomology) is known for them. They are special cases of a more general situation. To explain this, let us take, for example, the second equality. Then for $p = 7$ we let $\zeta = e^{2\pi i/7} \in \mathbb{C}$ be a primitive pth root of unity, and let

$$\omega = (-1 + \sqrt{-3})/2 = e^{2\pi i/3}$$

be a nontrivial $(p - 1)$th root of unity. Next take a generator of the multiplicative group $\mathbb{F}_p^\times$ (see Exercise 2 of §III.1); in our case $p = 7$ let us take 3. Then

$$g \underset{\text{def}}{=} \sum_{i=1}^{6} \omega^i \zeta^{3^i}$$

is known as a Gauss sum. It is not hard to verify that the right side of the second equality above is equal to $g^3/7$. More generally, one can prove that, whenever a/d is a rational number whose denominator divides $p - 1$, the *p*-adic number $\Gamma_p(a/d)^d$ is an element of the field $\mathbb{Q}(\omega)$, where ω is a primitive dth root of unity. (Exercise 10 gives another special case of this, in which $a/d = 1/2$, $\omega = -1$.) Namely, it turns out that $\Gamma_p(a/d)^d$ can be expressed in terms of suitable Gauss sums. (For a treatment of this, see Lang, *Cyclotomic Fields*, Vol. 2, or else Koblitz, *p-adic Analysis: a Short Course on Recent Work*.)

Note, by the way, that this shows a major difference between Γ_p and the classical Γ-function, since, for example, $\Gamma(1/3)$ is known to be transcendental.

13. Let $s = r/(p - 1)$ be a rational number in the interval $(0, 1)$, and let m be a positive integer not divisible by p. Prove that

$$\frac{\prod_{h=0}^{m-1} \Gamma_p((s + h)/m)}{\Gamma_p(s) \prod_{h=1}^{m-1} \Gamma_p(h/m)}$$

is equal to the Teichmüller representative of $m^{(1-s)(1-p)}$ (i.e., to the $(p - 1)$th root of unity in $\mathbb{Z}_p$ which is congruent mod p to $m^{1-p+r} \equiv m^r$). (Recall that if Γ_p is replaced by the classical Γ-function, then this expression equals m^{1-s}).

14. Prove that $\exp_p X$, $(\sin_p X)/X$, and $\cos_p X$ have no zeros in their regions of convergence, and that $E_p(X)$ has no zeros in $D(1^-)$.

15. Find the coefficients up through the X^4 term in $E_p(X)$ for $p = 2, 3$.

16. Find the coefficients in $E_p(X)$ through the X^{p-1} term. Find the coefficient of X^p. What fact from elementary number theory is reflected in the fact that the coefficient of X^p lies in $\mathbb{Z}_p$?

17. Use Dwork's lemma to give another proof that the coefficients of $E_p(X)$ are in $\mathbb{Z}_p$.

18. Use Dwork's lemma to prove: Let $f(X) = \exp(\sum_{i=0}^{\infty} b_i X^{p^i})$, $b_i \in \mathbb{Q}_p$. Then $f(X) \in 1 + X\mathbb{Z}_p[[X]]$ if and only if $b_{i-1} - pb_i \in p\mathbb{Z}_p$ for $i = 0, 1, 2, \ldots$ (where $b_{-1} \underset{\text{def}}{=} 0$).

3. Newton polygons for polynomials

Let $f(X) = 1 + \sum_{i=1}^{n} a_i X^i \in 1 + X\Omega[X]$ be a polynomial of degree n with coefficients in Ω and constant term 1. Consider the following sequence of points in the real coordinate plane:

$$(0, 0), (1, \operatorname{ord}_p a_1), (2, \operatorname{ord}_p a_2), \ldots, (i, \operatorname{ord}_p a_i), \ldots, (n, \operatorname{ord}_p a_n).$$

(If $a_i = 0$, we omit that point, or we think of it as lying "infinitely" far above the horizontal axis.) The *Newton polygon* of $f(X)$ is defined to be the "convex hull" of this set of points, i.e., the highest convex polygonal line joining $(0, 0)$ with $(n, \operatorname{ord}_p a_n)$ which passes on or below all of the points $(i, \operatorname{ord}_p a_i)$. Physically, this convex hull is constructed by taking a vertical line through $(0, 0)$ and rotating it about $(0, 0)$ counterclockwise until it hits any of the points $(i, \operatorname{ord}_p a_i)$, taking the segment joining $(0, 0)$ to the last such point $(i_1, \operatorname{ord}_p a_{i_1})$ that it hits as the first segment of the Newton polygon, then rotating the line further about $(i_1, \operatorname{ord}_p a_{i_1})$ until it hits a further point $(i, \operatorname{ord}_p a_i)$ $(i > i_1)$, taking the segment joining $(i_1, \operatorname{ord}_p a_{i_1})$ to the last such point $(i_2, \operatorname{ord}_p a_{i_2})$ as the second segment, then rotating the line about $(i_2, \operatorname{ord}_p a_{i_2})$ and so on, until you reach $(n, \operatorname{ord}_p a_n)$.

As an example, Figure 1 shows the Newton polygon for $f(X) = 1 + X^2 + \frac{1}{3}X^3 + 3X^4$ in $\mathbb{Q}_3[X]$.

By the vertices of the Newton polygon we mean the points $(i_j, \operatorname{ord}_p a_{i_j})$ where the slopes change. If a segment joins a point (i, m) to (i', m'), its slope is $(m' - m)/(i' - i)$; by the "length of the slope" we mean $i' - i$, i.e., the length of the projection of the corresponding segment onto the horizontal axis.

Lemma 4. *In the above notation, let $f(X) = (1 - X/\alpha_1)\cdots(1 - X/\alpha_n)$ be the factorization of $f(X)$ in terms of its roots $\alpha_i \in \Omega$. Let $\lambda_i = \operatorname{ord}_p 1/\alpha_i$. Then, if λ is a slope of the Newton polygon having length l, it follows that precisely l of the λ_i are equal to λ.*

97

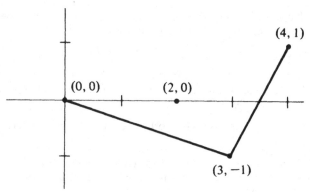

Figure IV.1

In other words, the slopes of the Newton polygon of $f(X)$ "are" (counting multiplicity) the p-adic ordinals of the reciprocal roots of $f(X)$.

PROOF. We may suppose the α_i to be arranged so that $\lambda_1 \le \lambda_2 \le \cdots \le \lambda_n$. Say $\lambda_1 = \lambda_2 = \cdots = \lambda_r < \lambda_{r+1}$. We first claim that the first segment of the Newton polygon is the segment joining $(0, 0)$ to $(r, r\lambda_1)$. Recall that each a_i is expressed in terms of $1/\alpha_1, 1/\alpha_2, \ldots, 1/\alpha_n$ as $(-1)^i$ times the ith symmetric polynomial, i.e., the sum of all possible products of i of the $1/\alpha$'s. Since the p-adic ordinal of such a product is at least $i\lambda_1$, the same is true for a_i. Thus, the point $(i, \mathrm{ord}_p\, a_i)$ is *on or above* the point $(i, i\lambda_1)$, i.e., on or above the line joining $(0, 0)$ to $(r, r\lambda_1)$.

Now consider a_r. Of the various products of r of the $1/\alpha$'s, exactly one has p-adic ordinal $r\lambda_1$, namely, the product $1/(\alpha_1\alpha_2 \cdots \alpha_r)$. All of the other products have p-adic ordinal $> r\lambda_1$, since we must include at least one of the $\lambda_{r+1}, \lambda_{r+2}, \ldots, \lambda_n$. Thus, a_r is a sum of something with ordinal $r\lambda_1$ and something with ordinal $> r\lambda_1$, so, by the "isosceles triangle principle," $\mathrm{ord}_p\, a_r = r\lambda_1$.

Now suppose $i > r$. In the same way as before, we see that all of the products of i of the $1/\alpha$'s have p-adic ordinal $> i\lambda_1$. Hence, $\mathrm{ord}_p\, a_i > i\lambda_1$. If we now think of how the Newton polygon is constructed, we see that we have shown that its first segment is the line joining $(0, 0)$ with $(r, r\lambda_1)$.

The proof that, if we have $\lambda_s < \lambda_{s+1} = \lambda_{s+2} = \cdots = \lambda_{s+r} < \lambda_{s+r+1}$, then the line joining $(s, \lambda_1 + \lambda_2 + \cdots + \lambda_s)$ to $(s + r, \lambda_1 + \lambda_2 + \cdots + \lambda_s + r\lambda_{s+1})$ is a segment of the Newton polygon, is completely analogous and will be left to the reader. $\qquad\square$

4. Newton polygons for power series

Now let $f(X) = 1 + \sum_{i=1}^{\infty} a_i X^i \in 1 + X\Omega[[X]]$ be a power series. Define $f_n(X) = 1 + \sum_{i=1}^{n} a_i X^i \in 1 + X\Omega[X]$ to be the nth partial sum of $f(X)$. In this section we suppose that $f(X)$ is not a polynomial, i.e., infinitely many

Figure IV.2

a_i are nonzero. The Newton polygon of $f(X)$ is defined to be the "limit" of the Newton polygons of the $f_n(X)$. More precisely, we follow the same recipe as in the construction of the Newton polygon of a polynomial: plot all of the points $(0, 0)$, $(1, \operatorname{ord}_p a_1)$, $\ldots$, $(i, \operatorname{ord}_p a_i)$, $\ldots$; rotate the vertical line through $(0, 0)$ until it hits a point $(i, \operatorname{ord}_p a_i)$, then rotate it about the farthest such point it hits, and so on. But we must be careful to notice that three things can happen:

(1) We get infinitely many segments of finite length. For example, take $f(X) = 1 + \sum_{i=1}^{\infty} p^{i^2} X^i$, whose Newton polygon is a polygonal line inscribed in the right half of the parabola $y = x^2$ (see Figure 2).

(2) At some point the line we're rotating simultaneously hits points $(i, \operatorname{ord}_p a_i)$ which are arbitrarily far out. In that case, the Newton polygon has a finite number of segments, the last one being infinitely long. For example, the Newton polygon of $f(X) = 1 + \sum_{i=1}^{\infty} X^i$ is simply one infinitely long horizontal segment.

(3) At some point the line we're rotating has not yet hit any of the $(i, \operatorname{ord}_p a_i)$ which are farther out, but, if we rotated it any farther at all, it would rotate past such points, i.e., it would pass above some of the $(i, \operatorname{ord}_p a_i)$. A simple example is $f(X) = 1 + \sum_{i=1}^{\infty} p X^i$. In that case, when the line through $(0, 0)$ has rotated to the horizontal position, it can rotate no farther without passing above some of the points $(i, 1)$. When this happens, we let the last segment of the Newton polygon have slope equal to the least upper bound of all possible slopes for which it passes below all of the $(i, \operatorname{ord}_p a_i)$. In our example, the slope is 0, and the Newton polygon consists of one infinite horizontal segment (see Figure 3).

A degenerate case of possibility (3) occurs when the vertical line through $(0, 0)$ cannot be rotated at all without crossing above some points $(i, \operatorname{ord}_p a_i)$. For example, this is what happens with $f(X) = \sum_{i=0}^{\infty} X^i/p^{i^2}$. In that case, $f(X)$

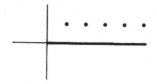

Figure IV.3

is easily seen to have zero radius of convergence, i.e., $f(x)$ diverges for any nonzero x. In what follows we shall exclude that case from consideration and shall suppose that $f(X)$ has a nontrivial disc of convergence.

In the case of polynomials, the Newton polygon is useful because it allows us to see at a glance at what radii the reciprocal roots are located. We shall prove that the Newton polygon of a power series $f(X)$ similarly tells us where the zeros of $f(X)$ lie. But first, let's make an ad hoc study of a particularly illustrative example.

Let

$$f(X) = 1 + \frac{X}{2} + \frac{X^2}{3} + \cdots + \frac{X^i}{i+1} + \cdots = -\frac{1}{X}\log_p(1-X).$$

The Newton polygon of $f(X)$ (see Figure 4, in which $p = 3$) is the polygonal line joining the points $(0, 0)$, $(p - 1, -1)$, $(p^2 - 1, -2), \ldots, (p^j - 1, -j), \ldots$; it is of type (1) in the list at the beginning of this section. If the power series analogue of Lemma 4 of §3 is to hold, we would expect from looking at this Newton polygon that $f(X)$ has precisely $p^{j+1} - p^j$ roots of p-adic ordinal $1/(p^{j+1} - p^j)$.

But what are the roots of $-1/X \log_p(1 - X)$? First, if $x = 1 - \zeta$, where ζ is a primitive p^{j+1}th root of 1, we know by Exercise 7 of §III.4 that $\operatorname{ord}_p x = 1/(p^{j+1} - p^j)$; and we know by the discussion of $\log_p$ in §IV.1 that $\log_p (1 - x) = \log_p \zeta = 0$. Since there are $p^{j+1} - p^j$ primitive p^{j+1}th roots of 1, this gives us all of the predicted roots. Are there any other zeros of $f(X)$ in $D(1^-)$?

Let $x \in D(1^-)$ be such a root. Then for any j, $x_j = 1 - (1 - x)^{p^j} \in D(1^-)$ is also a root since $\log_p(1 - x_j) = p^j \log_p(1 - x) = 0$. But for j sufficiently large, we have $x_j \in D(p^{-1/(p-1)-})$. For $x_j \in D(p^{-1/(p-1)-})$, we have $1 - x_j = \exp_p(\log_p(1 - x_j)) = \exp_p 0 = 1$. Hence $(1 - x)^{p^j} = 1$, and x must be one

Figure IV.4

of the roots we already considered. Thus, the appearance of the Newton polygon agrees with our knowledge of all of the roots of $\log_p(1 - X)$.

We now proceed to prove that the Newton polygon plays the same role for power series as for polynomials. But first we prove a much simpler result: that the radius of convergence of a power series can be seen at a glance from its Newton polygon.

Lemma 5. *Let b be the least upper bound of all slopes of the Newton polygon of* $f(X) = 1 + \sum_{i=1}^{\infty} a_i X^i \in 1 + X\Omega[[X]]$. *Then the radius of convergence is* p^b *(b may be infinite, in which case* $f(X)$ *converges on all of* Ω*).*

PROOF. First let $|x|_p < p^b$, i.e., $\mathrm{ord}_p x > -b$. Say $\mathrm{ord}_p x = -b'$, where $b' < b$. Then $\mathrm{ord}_p(a_i x^i) = \mathrm{ord}_p a_i - ib'$. But it is clear (see Figure 5) that, sufficiently far out, the $(i, \mathrm{ord}_p a_i)$ lie arbitrarily far above $(i, b'i)$, in other words, $\mathrm{ord}_p(a_i x^i) \to \infty$, and $f(X)$ converges at $X = x$.

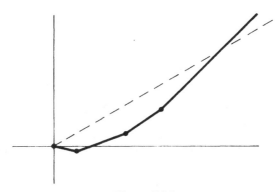

Figure IV.5

Now let $|x|_p > p^b$, i.e., $\mathrm{ord}_p x = -b' < -b$. Then we find in the same way that $\mathrm{ord}_p(a_i x^i) = \mathrm{ord}_p a_i - b'i$ is negative for infinitely many values of i. Thus $f(x)$ does not converge. We conclude that $f(X)$ has radius of convergence exactly p^b.   □

Remark. This lemma says nothing about convergence or divergence when $|x|_p = p^b$. It is easy to see that convergence *at* the radius of convergence ("on the circumference") can only occur in type (3) in the list at the beginning of this section, and then if and only if the distance that $(i, \mathrm{ord}_p a_i)$ lies above the last (infinite) segment approaches ∞ as $i \to \infty$. An example of this behavior is the power series $f(X) = 1 + \sum_{i=1}^{\infty} p^i X^{p^i}$, whose Newton polygon is the horizontal line extending from $(0, 0)$. This $f(X)$ converges when $\mathrm{ord}_p x = 0$.

One final remark should be made before beginning the proof of the power series analogue of Lemma 4. If $c \in \Omega$, $\mathrm{ord}_p c = \lambda$, and $g(X) = f(X/c)$, then the Newton polygon for g is obtained from that for f by subtracting the line

$y = \lambda x$—the line through $(0, 0)$ with slope λ—from the Newton polygon for f. This is because, if $f(X) = 1 + \sum a_i X^i$ and $g(X) = 1 + \sum b_i X^i$, then we have $\mathrm{ord}_p\, b_i = \mathrm{ord}_p(a_i/c^i) = \mathrm{ord}_p\, a_i - \lambda i$.

Lemma 6. *Suppose that λ_1 is the first slope of the Newton polygon of $f(X) = 1 + \sum_{i=1}^{\infty} a_i X^i \in 1 + X\Omega[[X]]$. Let $c \in \Omega$, $\mathrm{ord}_p\, c = \lambda \le \lambda_1$. Suppose that $f(X)$ converges on the closed disc $D(p^\lambda)$ (by Lemma 5, this automatically holds if $\lambda < \lambda_1$ or if the Newton polygon of $f(X)$ has more than one segment). Let*

$$g(X) = (1 - cX)f(X) \in 1 + X\Omega[[X]].$$

Then the Newton polygon of $g(X)$ is obtained by joining $(0, 0)$ to $(1, \lambda)$ and then translating the Newton polygon of $f(X)$ by 1 to the right and λ upward. In other words, the Newton polygon of $g(X)$ is obtained by "joining" the Newton polygon of the polynomial $(1 - cX)$ to the Newton polygon of the power series $f(X)$. In addition, if $f(X)$ has last slope λ_f and converges on $D(p^{\lambda_f})$, then $g(X)$ also converges on $D(p^{\lambda_f})$. Conversely, if $g(X)$ converges on $D(p^{\lambda_f})$, then so does $f(X)$.

PROOF. We first reduce to the special case $c = 1$, $\lambda = 0$. Suppose the lemma holds for that case, and we have $f(X)$ and $g(X)$ as in the lemma. Then $f_1(X) = f(X/c)$ and $g_1(X) = (1 - X)f_1(X)$ satisfy the conditions of the lemma with c, λ, λ_1 replaced by $1, 0, \lambda_1 - \lambda$, respectively (see the remark immediately preceding the statement of the lemma). Then the lemma, which we're assuming holds for f_1 and g_1, gives us the shape of the Newton polygon of $g_1(X)$ (and the convergence of g_1 on $D(p^{\lambda_f - \lambda})$ when f converges on $D(p^{\lambda_f})$). Since $g(X) = g_1(cX)$, if we again use the remark before the statement of the lemma, we obtain the desired information about the Newton polygon of $g(X)$. (See Figure 6.)

Thus, it suffices to prove Lemma 6 with $c = 1$, $\lambda = 0$. Let $g(X) = 1 + \sum_{i=1}^{\infty} b_i X^i$. Then, since $g(X) = (1 - X)f(X)$, we have $b_{i+1} = a_{i+1} - a_i$ for $i \ge 0$ (with $a_0 = 1$), and so

$$\mathrm{ord}_p\, b_{i+1} \ge \min(\mathrm{ord}_p\, a_{i+1}, \mathrm{ord}_p\, a_i),$$

with equality holding if $\mathrm{ord}_p\, a_{i+1} \ne \mathrm{ord}_p\, a_i$ (by the isosceles triangle principle). Since both $(i, \mathrm{ord}_p\, a_i)$ and $(i, \mathrm{ord}_p\, a_{i+1})$ lie on or above the Newton polygon of $f(X)$, so does $(i, \mathrm{ord}_p\, b_{i+1})$. If $(i, \mathrm{ord}_p\, a_i)$ is a vertex, then $\mathrm{ord}_p\, a_{i+1} > \mathrm{ord}_p\, a_i$, and so $\mathrm{ord}_p\, b_{i+1} = \mathrm{ord}_p\, a_i$. This implies that the Newton polygon of $g(X)$ must have the shape described in the lemma as far as the last vertex of the Newton polygon of $f(X)$. It remains to show that, in the case when the Newton polygon of $f(X)$ has a final infinite slope λ_f, $g(X)$ also does; and, if $f(X)$ converges on $D(p^{\lambda_f})$, then so does $g(X)$ (and conversely). Since $\mathrm{ord}_p\, b_{i+1} \ge \min(\mathrm{ord}_p\, a_{i+1}, \mathrm{ord}_p\, a_i)$, it immediately follows that $g(X)$ converges wherever $f(X)$ does. We must rule out the possibility that the Newton polygon of $g(X)$ has a slope λ_g which is *greater* than λ_f. If the Newton polygon of $g(X)$

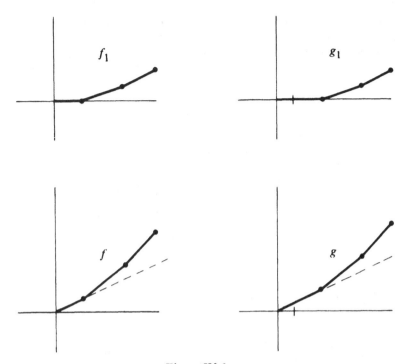

Figure IV.6

did have such a slope, then for some large i, the point $(i + 1, \text{ord}_p\, a_i)$ would lie below the Newton polygon of $g(X)$. Then we would have $\text{ord}_p\, b_j >$ $\text{ord}_p\, a_i$ for all $j \geq i + 1$. This first of all implies that $\text{ord}_p\, a_{i+1} = \text{ord}_p\, a_i$, because $a_{i+1} = b_{i+1} + a_i$; then in the same way $\text{ord}_p\, a_{i+2} = \text{ord}_p\, a_{i+1}$, and so on: $\text{ord}_p\, a_j = \text{ord}_p\, a_i$ for all $j > i$. But this contradicts the assumed convergence of $f(X)$ on $D(1)$. The converse assertion (convergence of g implies convergence of f) is proved in the same way. $\square$

Lemma 7. *Let $f(X) = 1 + \sum_{i=1}^{\infty} a_i X^i \in 1 + X\Omega[[X]]$ have Newton polygon with first slope λ_1. Suppose that $f(X)$ converges on the closed disc $D(p^{\lambda_1})$, and also suppose that the line through $(0, 0)$ with slope λ_1 actually passes through a point $(i, \text{ord}_p\, a_i)$. (Both of these conditions automatically hold if the Newton polygon has more than one slope.) Then there exists an x for which $\text{ord}_p\, x = -\lambda_1$ and $f(x) = 0$.*

PROOF. For simplicity, we first consider the case $\lambda_1 = 0$, and then reduce the general case to this one. In particular, $\text{ord}_p\, a_i \geq 0$ for all i and $\text{ord}_p\, a_i \to \infty$ as $i \to \infty$. Let $N \geq 1$ be the greatest i for which $\text{ord}_p\, a_i = 0$. (Except in the case when the Newton polygon of $f(X)$ is only one infinite horizontal line, this N is the length of the first segment, of slope $\lambda_1 = 0$.) Let $f_n(X) = 1 + \sum_{i=1}^{n} a_i X^i$. By Lemma 4, for $n \geq N$ the polynomial $f_n(X)$ has precisely N

roots $x_{n,1}, \ldots, x_{n,N}$ with $\operatorname{ord}_p x_{n,i} = 0$. Let $x_N = x_{N,1}$, and for $n \geq N$ let x_{n+1} be any of the $x_{n+1,1}, \ldots, x_{n+1,N}$ with $|x_{n+1,i} - x_n|_p$ minimal. We claim that $\{x_n\}$ is Cauchy, and that its limit x has the desired properties.

For $n \geq N$ let S_n denote the set of roots of $f_n(X)$ (counted with their multiplicities). Then for $n \geq N$ we have

$$|f_{n+1}(x_n) - f_n(x_n)|_p = |f_{n+1}(x_n)|_p \quad (\text{since } f_n(x_n) = 0)$$

$$= \prod_{x \in S_{n+1}} \left|1 - \frac{x_n}{x}\right|_p$$

$$= \prod_{i=1}^{N} |1 - x_n/x_{n+1,i}|_p \quad (\text{since if } x \in S_{n+1} \text{ has } \operatorname{ord}_p x < 0,$$

$$\text{we then have } |1 - x_n/x|_p = 1)$$

$$= \prod_{i=1}^{N} |x_{n+1,i} - x_n|_p \quad (\text{since } |x_{n+1,i}|_p = 1)$$

$$\geq |x_{n+1} - x_n|_p^N,$$

by the choice of x_{n+1}. Thus,

$$|x_{n+1} - x_n|_p^N \leq |f_{n+1}(x_n) - f_n(x_n)|_p = |a_{n+1} x_n^{n+1}|_p = |a_{n+1}|_p.$$

Since $|a_{n+1}|_p \to 0$ as $n \to \infty$, it follows that $\{x_n\}$ is Cauchy.

If $x_n \to x \in \Omega$, we further have $f(x) = \lim_{n \to \infty} f_n(x)$, while

$$|f_n(x)|_p = |f_n(x) - f_n(x_n)|_p = |x - x_n|_p \left| \sum_{i=1}^{n} a_i \frac{x^i - x_n^i}{x - x_n} \right|_p \leq |x - x_n|_p,$$

since $|a_i|_p \leq 1$ and $|(x^i - x_n^i)/(x - x_n)|_p = |x^{i-1} + x^{i-2} x_n + x^{i-3} x_n^2 + \cdots + x_n^{i-1}|_p \leq 1$. Hence, $f(x) = \lim_{n \to \infty} f_n(x) = 0$. This proves the lemma when $\lambda_1 = 0$.

Now the general case follows easily. Let $\pi \in \Omega$ be any number such that $\operatorname{ord}_p \pi = \lambda_1$. Note that such a π exists, for example, take an ith root of an a_i for which $(i, \operatorname{ord}_p a_i)$ lies on the line through $(0, 0)$ with slope λ_1. Now let $g(X) = f(X/\pi)$. Then $g(X)$ satisfies the conditions of the lemma with $\lambda_1 = 0$. So, by what's already been proved, there exists an x_0 with $\operatorname{ord}_p x_0 = 0$ and $g(x_0) = 0$. Let $x = x_0/\pi$. Then $\operatorname{ord}_p x = -\lambda_1$ and $f(x) = f(x_0/\pi) = g(x_0) = 0$. $\square$

Lemma 8. *Let* $f(X) = 1 + \sum_{i=1}^{\infty} a_i X^i \in 1 + X\Omega[[X]]$ *converge and have value* 0 *at* α. *Let* $g(X) = 1 + \sum_{i=1}^{\infty} b_i X^i$ *be obtained by dividing* $f(X)$ *by* $1 - X/\alpha$, *or equivalently, by multiplying* $f(X)$ *by the series* $1 + X/\alpha + X^2/\alpha^2 + \cdots + X^i/\alpha^i + \cdots$. *Then* $g(X)$ *converges on* $D(|\alpha|_p)$.

PROOF. Let $f_n(X) = 1 + \sum_{i=1}^{n} a_i X^i$. Clearly,

$$b_i = 1/\alpha^i + a_1/\alpha^{i-1} + a_2/\alpha^{i-2} + \cdots + a_{i-1}/\alpha + a_i,$$

so that

$$b_i \alpha^i = f_i(\alpha).$$

Hence $|b_i \alpha^i|_p = |f_i(\alpha)|_p \to 0$ as $i \to \infty$, because $f(\alpha) = 0$. $\square$

Theorem 14 (*p*-adic Weierstrass Preparation Theorem). *Let $f(X) = 1 + \sum_{i=1}^{\infty} a_i X^i \in 1 + X\Omega[[X]]$ converge on $D(p^\lambda)$. Let N be the total horizontal length of all segments of the Newton polygon having slope $\leq \lambda$ if this horizontal length is finite (i.e., if the Newton polygon of $f(X)$ does not have an infinitely long last segment of slope λ). If, on the other hand, the Newton polygon of $f(X)$ has last slope λ, let N be the greatest i such that $(i, \mathrm{ord}_p a_i)$ lies on that last segment (there must be a greatest such i, because $f(X)$ converges on $D(p^\lambda)$). Then there exists a polynomial $h(X) \in 1 + X\Omega[X]$ of degree N and a power series $g(X) = 1 + \sum_{i=1}^{\infty} b_i X^i$ which converges and is nonzero on $D(p^\lambda)$, such that*

$$h(X) = f(X) \cdot g(X).$$

The polynomial $h(X)$ is uniquely determined by these properties, and its Newton polygon coincides with the Newton polygon of $f(X)$ out to $(N, \mathrm{ord}_p a_N)$.

PROOF. We use induction on N. First suppose $N = 0$. Then we must show that $g(X)$, the inverse power series of $f(X)$, converges and is nonzero on $D(p^\lambda)$. This was part of Exercise 3 of §IV.1, but, since this is an important fact, we'll prove it here in case you skipped that exercise. As usual (see the proofs of Lemma 6 and 7 and the remark right before the statement of Lemma 6), we can easily reduce to the case $\lambda = 0$.

Thus, suppose $f(X) = 1 + \sum a_i X^i$, $\mathrm{ord}_p a_i > 0$, $\mathrm{ord}_p a_i \to \infty$, $g(X) = 1 + \sum b_i X^i$. Since $f(X)g(X) = 1$, we obtain

$$b_i = -(b_{i-1}a_1 + b_{i-2}a_2 + \cdots + b_1 a_{i-1} + a_i) \text{ for } i \geq 1,$$

from which it readily follows by induction on i that $\mathrm{ord}_p b_i > 0$. Next, we must show that $\mathrm{ord}_p b_i \to \infty$ as $i \to \infty$. Suppose we are given some large M. Choose m so that $i > m$ implies $\mathrm{ord}_p a_i > M$. Let

$$\varepsilon = \min(\mathrm{ord}_p a_1, \mathrm{ord}_p a_2, \ldots, \mathrm{ord}_p a_m) > 0.$$

We claim that $i > nm$ implies that $\mathrm{ord}_p b_i > \min(M, n\varepsilon)$, from which it will follow that $\mathrm{ord}_p b_i \to \infty$. We prove this claim by induction on n. It's trivial for $n = 0$. Suppose $n \geq 1$ and $i > nm$. We have

$$b_i = -(b_{i-1}a_1 + \cdots + b_{i-m}a_m + b_{i-(m+1)}a_{m+1} + \cdots + a_i).$$

The terms $b_{i-j}a_j$ with $j > m$ have $\mathrm{ord}_p(b_{i-j}a_j) \geq \mathrm{ord}_p a_j > M$, while the terms with $j \leq m$ have $\mathrm{ord}_p(b_{i-j}a_j) \geq \mathrm{ord}_p b_{i-j} + \varepsilon > \min(M, (n-1)\varepsilon) + \varepsilon$ by the induction assumption (since $i - j > (n-1)m$) and the definition of ε. Hence all summands in the expression for b_i have $\mathrm{ord}_p > \min(M, n\varepsilon)$. This proves the claim, and hence the theorem for $N = 0$.

Now suppose $N \geq 1$, and the theorem holds for $N - 1$. Let $\lambda_1 \leq \lambda$ be the first slope of the Newton polygon of $f(X)$. Using Lemma 7, we find an α such that $f(\alpha) = 0$ and $\text{ord}_p \alpha = -\lambda_1$. Let

$$f_1(X) = f(X)\left(1 + \frac{X}{\alpha} + \frac{X^2}{\alpha^2} + \cdots + \frac{X^i}{\alpha^i} + \cdots\right)$$

$$= 1 + \sum a_i' X^i \in 1 + X\Omega[[X]].$$

By Lemma 8, $f_1(X)$ converges on $D(p^{\lambda_1})$. Let $c = 1/\alpha$, so that: $f(X) = (1 - cX)f_1(X)$. If the Newton polygon of $f_1(X)$ had first slope λ_1' less than λ_1, it would follow by Lemma 7 that $f_1(X)$ has a root with p-adic ordinal $-\lambda_1'$, and then so would $f(X)$, which it is easy to check is impossible. Hence $\lambda_1' \geq \lambda_1$, and we have the conditions of Lemma 6 (with f_1, f, λ_1', and λ_1 playing the roles of f, g, λ_1, and λ, respectively). Lemma 6 then tells us that $f_1(X)$ has the same Newton polygon as $f(X)$, minus the segment from $(0, 0)$ to $(1, \lambda_1)$. In addition, in the case when f (and hence f_1) have last slope λ, because f converges on $D(p^\lambda)$, Lemma 6 further tells us that f_1 must also converge on $D(p^\lambda)$.

Thus, $f_1(X)$ satisfies the conditions of the theorem with N replaced by $N - 1$. By the induction assumption, we can find an $h_1(X) \in 1 + X\Omega[X]$ of degree $N - 1$ and a series $g(X) \in 1 + X\Omega[[X]]$ which converges and is nonzero on $D(p^\lambda)$, such that

$$h_1(X) = f_1(X) \cdot g(X).$$

Then, multiplying both sides by $(1 - cX)$ and setting $h(X) = (1 - cX)h_1(X)$, we have

$$h(X) = f(X) \cdot g(X),$$

with $h(X)$ and $g(X)$ having the required properties.

Finally, suppose that $\tilde{h}(X) \in 1 + X\Omega[X]$ is another polynomial of degree N such that $\tilde{h}(X) = f(X)g_1(X)$, where $g_1(X)$ converges and is nonzero on $D(p^\lambda)$. Since $\tilde{h}(X)g(X) = f(X)g(X)g_1(X) = h(X)g_1(X)$, uniqueness of $h(X)$ follows if we prove the claim: $\tilde{h}g = hg_1$ implies that $\tilde{h}$ and h have the same zeros with the same multiplicities. This can be shown by induction on N. For $N = 1$ it is obvious, because $\tilde{h}(x) = 0 \Leftrightarrow h(x) = 0$ for $x \in D(p^\lambda)$. Now suppose $N > 1$. Without loss of generality we may assume that $-\lambda$ is ord_p of a root α of $h(X)$ having minimal ord_p. Since α is a root of both $h(X)$ and $\tilde{h}(X)$ of minimal ord_p, we can divide both sides of the equality $\tilde{h}(X)g(X) = h(X)g_1(X)$ by $(1 - X/\alpha)$, using Lemma 8, and thereby reduce to the case of our claim with N replaced by $N - 1$. This completes the proof of Theorem 14. $\square$

Corollary. *If a segment of the Newton polygon of $f(X) \in 1 + X\Omega[[X]]$ has finite length N and slope λ, then there are precisely N values of x counting multiplicity for which $f(x) = 0$ and $\text{ord}_p x = -\lambda$.*

Another consequence of Theorem 14 is that a power series which converges everywhere factors into the (infinite) product of $(1 - X/r)$ over all of its roots r, and, in particular, if it converges everywhere and has no zeros, it must be a constant. (See Exercise 13 below.) This contrasts with the real or complex case, where we have the function e^x (or, more generally, $e^{h(x)}$, where h is any everywhere convergent power series). In complex analysis, the analogous infinite product expansion of an everywhere convergent power series in terms of its roots is more complicated than in the p-adic case; exponential factors have to be thrown in to obtain the "Weierstrass product" of an "entire" function of a complex variable.

Thus, the simple infinite product expansion that results from Theorem 14 in the p-adic case is possible thanks to the absence of an everywhere convergent exponential function. So in the present context we're lucky that $\exp_p$ has bad convergence. But in other contexts—for example, p-adic differential equations—the absence of a nicely convergent exp makes life very complicated.

EXERCISES

1. Find the Newton polygon of the following polynomials:
   (i) $1 - X + pX^2$ (ii) $1 - X^3/p^2$ (iii) $1 + X^2 + pX^4 + p^3X^6$
   (iv) $\sum_{i=1}^{p} iX^{i-1}$ (v) $(1 - X)(1 - pX)(1 - p^3X)$ (do this in two ways)
   (vi) $\prod_{i=1}^{p^2}(1 - iX)$.

2. (a) Let $f(X) \in 1 + X\mathbb{Z}_p[X]$ have Newton polygon consisting of one segment joining $(0, 0)$ to the point (n, m). Show that if n and m are relatively prime, then $f(X)$ cannot be factored as a product of two polynomials with coefficients in $\mathbb{Z}_p$.
   (b) Use part (a) to give another proof of the Eisenstein irreducibility criterion (see Exercise 14 of §I.5).
   (c) Is the converse to (a) true or false, i.e., do all irreducible polynomials have Newton polygon of this type (proof or counterexample)?

3. Let $f(X) \in 1 + X\mathbb{Z}_p[X]$ be a polynomial of degree $2n$. Suppose you know that, whenever α is a reciprocal root of $f(X)$, so is p/α (with the same multiplicity). What does this tell you about the shape of the Newton polygon? Draw all possible shapes of Newton polygons of such $f(X)$ when $n = 1, 2, 3, 4$.

4. Find the Newton polygon of the following power series:
   (i) $\sum_{i=0}^{\infty} X^{p^i-1}/p^i$ (ii) $\sum_{i=0}^{\infty}((pX)^i + X^{p^i})$ (iii) $\sum_{i=0}^{\infty} i!X^i$
   (iv) $\sum_{i=0}^{\infty} X^i/i!$ (v) $(1 - pX^2)/(1 - p^2X^2)$ (vi) $(1 - p^2X)/(1 - pX)$
   (vii) $\prod_{i=0}^{\infty}(1 - p^iX)$ (viii) $\sum_{i=0}^{\infty} p^{[i\sqrt{2}]}X^i$.

5. Show that the slopes of the finite segments of the Newton polygon of a power series are rational numbers, but that the slope of the infinite segment (if there is one) need not be (give an example).

6. Show by a counterexample that Lemma 7 is false if we omit the condition that the line through $(0, 0)$ with slope λ_1 pass through a point $(i, \mathrm{ord}_p\, a_i)$, $i > 0$.

7. Suppose $f(X) \in 1 + X\Omega[[X]]$ has Newton polygon which is the degenerate case (3), i.e., a vertical line through (0, 0). In other words, if the vertical line through (0, 0) is rotated counterclockwise at all, it passes above some points $(i, \text{ord}_p \, a_i)$. Prove that $f(x)$ diverges for any nonzero $x \in \Omega$.

8. Let $f(X) = 1 + \sum_{i=1}^{\infty} a_i X^i \in 1 + X\Omega[[X]]$ converge in $D(p^\lambda)$ where λ is a rational number. Prove that $\max_{x \in D(p^\lambda)} |f(x)|_p$ is reached when $|x|_p = p^\lambda$, i.e., on the "circumference," and that this maximal value of $f(x)$ has p-adic ordinal equal to

$$\min_{i = 0, 1, \ldots}(\text{ord}_p \, a_i - i\lambda),$$

i.e., the minimum distance (which may be negative) of the point $(i, \text{ord}_p \, a_i)$ above the line through (0, 0) with slope λ.

9. Let $f(X) = \sum_{i=0}^{\infty} a_i X^i \in \mathbb{Z}_p[[X]]$. Suppose that $f(X)$ converges in the *closed* unit disc $D(1)$. Further suppose that at least two of the a_i are not divisible by p. Prove that $f(X)$ has a zero in $D(1)$.

10. Let $f(X)$ be a power series which converges on $D(r)$ and has an infinite number of zeros in $D(r)$. Show that $f(X)$ is identically zero.

11. Prove that $E_p(X)$ converges only in $D(1^-)$ (i.e., not in $D(1)$).

12. Let $g(X) = h(X)/f(X)$, where $g(X) \in 1 + X\Omega[[X]]$ has all coefficients in $D(1)$, and where $h(X)$ and $f(X) \in 1 + X\Omega[X]$ are polynomials with no common roots. Prove that $h(X)$ and $f(X)$ also have all coefficients in $D(1)$.

13. Suppose that $f(X) \in 1 + X\Omega[[X]]$ converges on all of Ω. For every λ, let $h_\lambda(X)$ be the $h(X)$ in Theorem 14. Prove that $h_\lambda \to f$ as $\lambda \to \infty$ (i.e., each coefficient of h_λ approaches the corresponding coefficient of f). Prove that f has infinitely many zeros if it is not a polynomial (but only a countably infinite set of zeros $r_1, r_2, \ldots$), and that $f(X) = \prod_{i=1}^{\infty} (1 - X/r_i)$. In particular, there is no nonconstant power series which converges and is nonzero everywhere (in contrast to the real or complex case, where, whenever $h(X)$ is *any* everywhere convergent power series, the power series $e^{h(X)}$ is an everywhere convergent and nonzero power series).

CHAPTER V

Rationality of the zeta-function of a set of equations over a finite field

1. Hypersurfaces and their zeta-functions

If F is a field, let $\mathbb{A}_F^n$ denote "n-dimensional affine space over F," i.e., the set of ordered n-tuples $(x_1, \ldots, x_n)$ of elements x_i of F. Let $f(X_1, \ldots, X_n) \in F[X_1, \ldots, X_n]$ be a polynomial in the n variables $X_1, \ldots, X_n$. By the *affine hypersurface* defined by f in $\mathbb{A}_F^n$, we mean

$$H_f \underset{\text{def}}{=} \{(x_1, \ldots, x_n) \in \mathbb{A}_F^n \mid f(x_1, \ldots, x_n) = 0\}.$$

The number $n - 1$ is called the *dimension* of H_f. We call H_f an *affine curve* if $n = 2$, i.e., if H_f is one-dimensional.

The companion concept to affine space is *projective space*. By n-dimensional projective space over F, denoted $\mathbb{P}_F^n$, we mean the set of equivalence classes of elements of

$$\mathbb{A}_F^{n+1} - \{(0, 0, \ldots, 0)\}$$

with respect to the equivalence relation

$$(x_0, x_1, \ldots, x_n) \sim (x_0', x_1', \ldots, x_n') \Leftrightarrow \exists \lambda \in F^\times \quad \text{with } x_i' = \lambda x_i, i = 0, \ldots, n.$$

In other words, as a set $\mathbb{P}_F^n$ is the set of all lines through the origin in $\mathbb{A}_F^{n+1}$.

$\mathbb{A}_F^n$ can be included in $\mathbb{P}_F^n$ by the map $(x_1, \ldots, x_n) \mapsto (1, x_1, \ldots, x_n)$. The image of $\mathbb{A}_F^n$ consists of all of $\mathbb{P}_F^n$ except for the "hyperplane at infinity" consisting of all equivalence classes of $(n + 1)$-tuples with zero x_0-coordinate. That hyperplane looks like a copy of $\mathbb{P}_F^{n-1}$, by virtue of the one-to-one correspondence

equiv. class of $(0, x_1, \ldots, x_n) \mapsto$ equiv. class of $(x_1, \ldots, x_n)$.

(For example, if $n = 2$, the projective plane $\mathbb{P}_F^2$ can be thought of as the affine plane plus the "line at infinity.") Continuing in this way, we can write $\mathbb{P}_F^n$ as a disjoint union

$$\mathbb{A}_F^n \bigcup \mathbb{A}_F^{n-1} \bigcup \mathbb{A}_F^{n-2} \bigcup \ldots \bigcup \mathbb{A}_F^1 \bigcup \text{point.}$$

By a *homogeneous* polynomial $\tilde{f}(X_0, \ldots, X_n) \in F[X_0, \ldots, X_n]$ of degree d we mean a linear combination of monomials of the *same total degree d*. For example, $X_0{}^3 + X_0{}^2 X_1 - 3X_1 X_2 X_3 + X_3{}^3$ is homogeneous of degree 3. Given a polynomial $f(X_1, \ldots, X_n) \in F[X_1, \ldots, X_n]$ of degree d, its *homogeneous completion* $\tilde{f}(X_0, X_1, \ldots, X_n)$ is the polynomial

$$X_0^d f(X_1/X_0, \ldots, X_n/X_0),$$

which is clearly homogeneous of degree d. For example, the homogeneous completion of $X_3{}^3 - 3X_1 X_2 X_3 + X_1 + 1$ is the above example of a homogeneous polynomial of degree 3.

If $\tilde{f}(X_0, \ldots, X_n)$ is homogeneous, and if $\tilde{f}(x_0, \ldots, x_n) = 0$, then also $\tilde{f}(\lambda x_0, \ldots, \lambda x_n) = 0$ for $\lambda \in F^\times$. Hence it makes sense to talk of the set of points (equivalence classes of $(n+1)$-tuples) of $\mathbb{P}_F^n$ at which $\tilde{f}$ vanishes. That set of points $\tilde{H}_{\tilde{f}}$ is called the *projective hypersurface* defined by $\tilde{f}$ in $\mathbb{P}_F^n$.

If $\tilde{f}(X_0, \ldots, X_n)$ is the homogeneous completion of $f(X_1, \ldots, X_n)$, then $\tilde{H}_{\tilde{f}}$ is called the *projective completion* of H_f. Intuitively, $\tilde{H}_{\tilde{f}}$ is obtained from H_f by "throwing in the points at infinity toward which H_f is heading." For example, if H_f is the hyperbola (say $F = \mathbb{R}$)

$$\frac{X_1{}^2}{a^2} - \frac{X_2{}^2}{b^2} = 1,$$

then $\tilde{f}(X_0, X_1, X_2) = X_1{}^2/a^2 - X_2{}^2/b^2 - X_0{}^2$, and $\tilde{H}_{\tilde{f}}$ consists of

$$\{(1, X_1, X_2) \mid X_1{}^2/a^2 - X_2{}^2/b^2 = 1\} \cup \{(0, 1, X_2) \mid X_2 = \pm b/a\},$$

i.e., H_f plus the points on the line at infinity corresponding to the slopes of the asymptotes.

Now let K be any field containing F. If the coefficients of a polynomial are in F, then they are also in K, so we may consider the "K-points" of H_f, i.e.,

$$H_f(K) \underset{\text{def}}{=} \{(x_1, \ldots, x_n) \in \mathbb{A}_K^n \mid f(x_1, \ldots, x_n) = 0\}.$$

If $\tilde{f}(X_0, \ldots, X_n)$ is homogeneous, we similarly define $\tilde{H}_{\tilde{f}}(K)$.

We shall be working with finite fields $F = \mathbb{F}_q$ and finite field extensions $K = \mathbb{F}_{q^s}$. In that case $H_f(K)$ and $\tilde{H}_{\tilde{f}}(K)$ consist of finitely many points, since there are only finitely many (namely, q^{sn}) n-tuples in all of $\mathbb{A}_K^n$ (and only finitely many points in $\mathbb{P}_K^n$). In what follows H_f (or $\tilde{H}_{\tilde{f}}$) will be fixed throughout the discussion. In that case we define the sequence $N_1, N_2, N_3, \ldots$ to be the number of $\mathbb{F}_q$-, $\mathbb{F}_{q^2}$-, $\mathbb{F}_{q^3}$-, $\ldots$ -points of H_f (or $\tilde{H}_{\tilde{f}}$), i.e.,

$$N_s \underset{\text{def}}{=} \#(H_f(\mathbb{F}_{q^s})).$$

Given any sequence of integers such as $\{N_s\}$ which has geometric or number theoretic significance, we can form the so-called "generating function" which captures all the information conveyed by the sequence $\{N_s\}$ in a power series. This is the "zeta-function," which is defined as the formal power series

$$\exp\left(\sum_{s=1}^{\infty} N_s T^s/s\right) \in \mathbb{Q}[[T]].$$

We write this function as $Z(H_f/\mathbb{F}_q; T)$, where $\mathbb{F}_q$ indicates what the original field F was. Note that the power series $Z(H_f/\mathbb{F}_q; T)$ has constant term 1.

Before giving some examples, we prove a couple of elementary lemmas.

Lemma 1. $Z(H_f/\mathbb{F}_q; T)$ *has coefficients in* $\mathbb{Z}$.

PROOF. We consider the K-points $P = (x_1, \ldots, x_n)$ of H_f (K a finite extension of $\mathbb{F}_q$) according to the least $s = s_0$ for which all $x_i \in \mathbb{F}_{q^{s_0}}$. If $P_j = (x_{1j}, \ldots, x_{nj})$, $j = 1, \ldots, s_0$, are the "conjugates" of P, i.e., $x_{i1}, \ldots, x_{is_0}$ are the conjugates of $x_i = x_{i1}$ over $\mathbb{F}_q$, then the P_j are distinct, because if all of the x_i are left fixed by an automorphism σ of $\mathbb{F}_{q^{s_0}}$ over $\mathbb{F}_q$, it follows that they are all in a smaller field (namely, the "fixed field" of σ: $\{x \in \mathbb{F}_{q^{s_0}} \mid \sigma(x) = x\}$).

Now let's count the contribution of $P_1, \ldots, P_{s_0}$ to $Z(H_f/\mathbb{F}_q; T)$. Each of these points is an $\mathbb{F}_{q^s}$-point of H_f precisely when $\mathbb{F}_{q^s} \supset \mathbb{F}_{q^{s_0}}$, i.e., when $s_0 | s$ (see Exercise 1 of §III.1). Thus, these points contribute s_0 to $N_{s_0}, N_{2s_0}, N_{3s_0}, \ldots$, and so their contribution to $Z(H_f/\mathbb{F}_q; T)$ is:

$$\exp\left(\sum_{j=1}^{\infty} s_0 T^{js_0}/js_0\right) = \exp(-\log(1 - T^{s_0})) = \frac{1}{1 - T^{s_0}} = \sum_{j=0}^{\infty} T^{js_0}.$$

The whole zeta-function is then a product of series of this type (only finitely many of which has first T-term with degree $\leq s_0$), and so has integer coefficients. $\qquad\square$

Remark. Note that a corollary of the proof is that the coefficients are *nonnegative* integers.

Lemma 2. *The coefficient of* T^j *in* $Z(H_f/\mathbb{F}_q; T)$ *is* $\leq q^{nj}$.

PROOF. The maximum value for N_s is $q^{ns} = \# \mathbb{A}^n_{\mathbb{F}_{q^s}}$. The coefficients of $Z(H_f/\mathbb{F}_q; T)$ are clearly less than or equal to the coefficients of the series with N_s replaced by q^{ns}. But

$$\exp\left(\sum_{s=1}^{\infty} q^{ns}T^s/s\right) = \exp(-\log(1 - q^nT)) = 1/(1 - q^nT) = \sum_{j=0}^{\infty} q^{nj}T^j. \quad\square$$

As a simple example, let's compute the zeta-function of an affine line $L = H_{x_1} \subset \mathbb{A}^2_{\mathbb{F}_q}$. We have $N_s = q^s$, and so

$$Z(L/\mathbb{F}_q; T) = \exp(\textstyle\sum q^sT^s/s) = \exp(-\log(1 - qT)) = \frac{1}{1 - qT}.$$

The zeta-function is defined analogously for projective hypersurfaces, where now we use

$$\tilde{N}_s \underset{\text{def}}{=} \#(\tilde{H}_{\bar{f}}(\mathbb{F}_{q^s})).$$

For example, for a projective line $\tilde{L}$ we have $\tilde{N}_s = q^s + 1$, and so

$$Z(\tilde{L}/\mathbb{F}_q; T) = \exp(\textstyle\sum(q^sT^s/s + T^s/s))$$

$$= \exp(-\log(1 - qT) - \log(1 - T)) = \frac{1}{(1 - T)(1 - qT)}.$$

It turns out that it's much more natural to work with projective hypersurfaces than with affine ones.

For example, take the unit circle $X_1^2 + X_2^2 = 1$, whose projective completion is $\tilde{H}_{\tilde{f}}, \tilde{f} = X_1^2 + X_2^2 - X_0^2$. It's easier to compute $Z(\tilde{H}_{\tilde{f}}/\mathbb{F}_q; T)$ than $Z(\tilde{H}_f/\mathbb{F}_q; T)$. (We're assuming that $p = \mathrm{char}\ \mathbb{F}_q \neq 2$.) Why is it easier? Because there is a one-to-one correspondence between $\tilde{H}_{\tilde{f}}(K)$ and $\tilde{L}(K)$ ($\tilde{L}$ denotes the projective line). To construct this map, project from the south pole onto the line $X_2 = 1$, as shown in Figure 1. A simple computation gives:

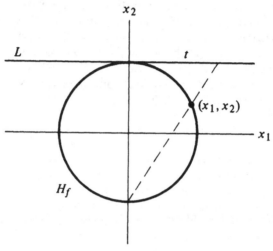

Figure V.1

$x_1 = 4t/(4 + t^2)$, $x_2 = (4 - t^2)/(4 + t^2)$, $t = 2x_1/(x_2 + 1)$. This map goes bad in the t-to-x direction if $t^2 = -4$, i.e., for 2 values of t if $q^s \equiv 1 \pmod 4$ and no values of t if $q^s \equiv 3 \pmod 4$ (see Exercise 8 of §III.1). It goes bad in the other direction when $x_2 = -1$, $x_1 = 0$. But if we take the projective completions and let (X_0, X_1, X_2) be coordinates for the completed circle and (X_0', X_1') for the completed line, then it is easy to check that we have a perfectly nice one-to-one correspondence given by

$$(x_0', x_1') \mapsto (4x_0'^2 + x_1'^2, 4x_0'x_1', 4x_0'^2 - x_1'^2);$$

$(x_0, x_1, x_2) \mapsto (x_2 + x_0, 2x_1)$ if $(x_2 + x_0, 2x_1) \neq (0, 0)$, and $(0, 1)$ otherwise.

The reader should carefully verify that this does in fact give a one-to-one correspondence between the projective line and the set of equivalence classes of triples (x_0, x_1, x_2) satisfying $x_1^2 + x_2^2 - x_0^2 = 0$. Thus, since $\tilde{N}_s$ is the same for $\tilde{H}_{\tilde{f}}$ and $\tilde{L}$, we have

$$Z(\tilde{H}_{\tilde{f}}/\mathbb{F}_q; T) = Z(\tilde{L}/\mathbb{F}_q; T) = 1/[(1 - T)(1 - qT)].$$

If we wanted to know $Z(H_f/\mathbb{F}_q; T)$, $f = X_1^2 + X_2^2 - 1$, we'd have to subtract from $\tilde{N}_s$ the points "at infinity" on $\tilde{H}_{\tilde{f}}$, i.e., those for which $x_1^2 +$

$x_2{}^2 = x_0{}^2$ and $x_0 = 0$. There are 2 such points whenever -1 has a square root in $\mathbb{F}_{q^s}$ and no such points otherwise.

Case (1). $q \equiv 1 \pmod 4$. Then -1 always has a square root in $\mathbb{F}_{q^s}$ (see Exercise 8 of §III.1), and

$$N_s = \tilde{N}_s - 2,$$

$$Z(H_f/\mathbb{F}_q; T) = \frac{Z(\tilde{H}_f/\mathbb{F}_q; T)}{\exp\left(\sum_{s=1}^{\infty} 2T^s/s\right)} = \frac{1/[(1-T)(1-qT)]}{1/(1-T)^2} = \frac{1-T}{1-qT}.$$

Case (2). $q \equiv 3 \pmod 4$. Then it is easy to see that $N_s = \tilde{N}_s$ if s is odd, and $N_s = \tilde{N}_s - 2$ if s is even, so that

$$Z(H_f/\mathbb{F}_q; T) = \frac{Z(\tilde{H}_f/\mathbb{F}_q; T)}{\exp\left(\sum_{s=1}^{\infty} 2T^{2s}/2s\right)} = \frac{1/[(1-T)(1-qT)]}{1/(1-T^2)} = \frac{1+T}{1-qT}.$$

Notice that in all of these examples, as well as in the examples in the exercises below, the zeta-function turns out to be a rational function, a ratio of polynomials. This is an important general fact, which Dwork first proved in 1960 using an ingenious application of p-adic analysis.

Theorem (Dwork). *The zeta-function of any affine* (or projective—see Exercise 5 below) *hypersurface is a ratio of two polynomials with coefficients in* $\mathbb{Q}$ (actually, with coefficients in $\mathbb{Z}$ and constant term 1—see Exercise 13 below).

The rest of this chapter will be devoted to Dwork's proof of this theorem.

We note that zeta-functions of hypersurfaces can be generalized to a broader class of objects, including affine or projective "algebraic varieties," which are the same as hypersurfaces except that they may be defined by more than one simultaneous polynomial equation. Dwork's theorem holds for algebraic varieties (see Exercise 4 below).

Dwork's theorem has profound practical implications for solving systems of polynomial equations over finite fields. It implies that there exists a finite set of complex numbers $\alpha_1, \ldots, \alpha_t, \beta_1, \ldots, \beta_u$ such that for *all* $s = 1, 2, 3, \ldots$ we have $N_s = \sum_{i=1}^{t} \alpha_i{}^s - \sum_{i=1}^{u} \beta_i{}^s$ (see Exercise 6 below). In other words, once we determine a finite set of data (the α_i and β_i)—and this data is already determined by a finite number of N_s—we'll have a simple formula which predicts *all* the remaining N_s. Admittedly, in order to really work with this in practice, we must know a bound on the degree of the numerator and denominator of our rational function $Z(H/\mathbb{F}_q; T)$ (see Exercises 7–9 below for more details). Actually, in all important cases the degree of the numerator and denominator, along with much additional information, is now known about the zeta-function. This information is contained in the famous Weil Conjectures (now proved, but whose proof, even in the simplest cases, goes well beyond the scope of this book).

V Rationality of the zeta-function of a set of equations over a finite field

The rationality of the zeta-function was part of this series of conjectures announced by A. Weil in 1949. Dwork's proof of rationality was the first major step toward the proof of the Weil Conjectures. The final step, Deligne's proof in 1973 of the so-called "Riemann Hypothesis" for algebraic varieties, was the culmination of a quarter century of intense research on the conjectures.

In the case of a projective hypersurface $\tilde{H}_{\tilde{f}}$ in n-dimensional space that is smooth (i.e., for which the partial derivatives of $\tilde{f}$ with respect to all the variables never vanish simultaneously), the Weil Conjectures say that the following:

(i) $Z(\tilde{H}_{\tilde{f}}/\mathbb{F}_q: T) = P(T)^{\pm 1}/((1 - T)(1 - qT)(1 - q^2T) \cdots (1 - q^{n-1}T))$, where $P(T) \in 1 + T\mathbb{Z}[T]$ has degree β, where β is a number related to the "topology" of the hypersurface (called its "Betti number;" when $\tilde{H}_{\tilde{f}}$ is a curve, this is twice the genus, or "number of handles," of the corresponding Riemann surface). Here the ± 1 means we take $P(T)$ if n is even and $1/P(T)$ if n is odd.

(ii) If α is a reciprocal root of $P(T)$, then so is q^{n-1}/α.

(iii) The complex absolute value of each of the reciprocal roots of $P(T)$ is $q^{(n-1)/2}$. (This is called the "Riemann Hypothesis" part of Weil's conjectures, out of analogy with the classical Riemann Hypothesis for the Riemann zeta-function—see Exercise 15 below.)

EXERCISES

1. What is the zeta-function of a point? What is $Z(\mathbb{A}^n_{\mathbb{F}_q}/\mathbb{F}_q; T)$?

2. Compute $Z(\mathbb{P}^n_{\mathbb{F}_q}/\mathbb{F}_q; T)$.

3. Let $f(X_1, \ldots, X_n) = X_n + g(X_1, \ldots, X_{n-1})$, where $g \in \mathbb{F}_q[X_1, \ldots, X_{n-1}]$. Prove that

$$Z(H_f/\mathbb{F}_q; T) = Z(\mathbb{A}^{n-1}_{\mathbb{F}_q}/\mathbb{F}_q; T).$$

4. Let $f_1(X_1, \ldots, X_n), f_2(X_1, \ldots, X_n), \ldots, f_r(X_1, \ldots, X_n) \in \mathbb{F}_q[X_1, \ldots, X_n]$, and let $H_{\{f_1, f_2, \ldots, f_r\}}(\mathbb{F}_{q^s}) \subset \mathbb{A}^n_{\mathbb{F}_{q^s}}$ be the set of n-tuples of elements of $\mathbb{F}_{q^s}$ which satisfy *all* of the equations $f_i = 0$, $i = 1, 2, \ldots, r$:

$$H_{\{f_1, f_2, \ldots, f_r\}}(\mathbb{F}_{q^s}) \underset{\text{def}}{=} \{(x_1, \ldots, x_n) \in \mathbb{A}^n_{\mathbb{F}_{q^s}} \mid f_1(x_1, \ldots, x_n) = \cdots$$
$$= f_r(x_1, \ldots, x_n) = 0\}.$$

Such an H is called an (affine) algebraic variety. Let $N_s = \#H(\mathbb{F}_{q^s})$ (where H is short for $H_{\{f_1, \ldots, f_r\}}$), and define the zeta-function as before: $Z(H/\mathbb{F}_q; T) \underset{\text{def}}{=} \exp(\sum_{s=1}^{\infty} N_s T^s/s)$. Prove that Dwork's theorem for affine hypersurfaces implies Dwork's theorem for affine algebraic varieties.

5. Prove that if Dwork's theorem holds for affine hypersurfaces, then it holds for projective hypersurfaces.

6. Prove that Dwork's theorem is equivalent to: There exist algebraic complex numbers $\alpha_1, \ldots, \alpha_t, \beta_1, \ldots, \beta_u$, such that each conjugate of an α is an α, each conjugate of an α is an α, each conjugate of a β is a β, and we have:

$$N_s = \sum_{i=1}^{t} \alpha_i{}^s - \sum_{i=1}^{u} \beta_i{}^s, \quad \text{for all } s = 1, 2, 3, \ldots.$$

7. It is known that the zeta-function of a smooth cubic projective curve $\tilde{E} = \tilde{H}_{\tilde{f}}$ (thus, dim $\tilde{E} = 1$, deg $\tilde{f} = 3$; $\tilde{E}$ is called an "elliptic curve") is always of the form: $(1 + aT + qT^2)/[(1 - T)(1 - qT)]$ for some $a \in \mathbb{Z}$. Show that if you know the number of points in $\tilde{E}(\mathbb{F}_q)$, you can determine: (1) a, and (2) $\#\tilde{E}(\mathbb{F}_{q^s})$ for any s.

8. Using the fact stated in the previous problem, find $Z(\tilde{H}_{\tilde{f}}/\mathbb{F}_q; T)$ when $f(X_1, X_2)$ is:

   (i) $X_2{}^3 - X_1{}^3 - 1$ and $q \equiv 2 \pmod 3$;
   (ii) $X_2{}^2 - X_1{}^3 + X_1$ and $q \equiv 3 \pmod 4$, also $q = 5, 13,$ and 9.

9. Suppose we know that $Z(H/\mathbb{F}_q; T)$ is a rational function whose numerator has degree m and whose denominator has degree n. Prove that $N_s = \#H(\mathbb{F}_{q^s})$ for $s = 1, 2, 3, \ldots, m + n$ uniquely determine all of the other N_s.

10. Compute $Z(H_f/\mathbb{F}_q; T)$ when H_f is the 3-dimensional hypersurface defined by

$$X_1 X_4 - X_2 X_3 = 1.$$

11. Compute $Z(H_f/\mathbb{F}_q; T)$ and $Z(\tilde{H}_{\tilde{f}}/\mathbb{F}_q; T)$ ($\tilde{f}$ = homogeneous completion of f) when H_f is the curve:

   (i) $X_1 X_2 = 0$ (ii) $X_1 X_2 (X_1 + X_2 + 1) = 0$ (iii) $X_2{}^2 - X_1{}^2 = 1$
   (iv) $X_2{}^2 = X_1{}^3$ (v) $X_2{}^2 = X_1{}^3 + X_1{}^2$.

12. Lines in $\mathbb{P}^3$ are obtained by intersecting two distinct hyperplanes, i.e., a line is the set of equivalence classes of quadruples which satisfy simultaneously two given linear homogeneous polynomials. Let N_s be the number of lines in $\mathbb{P}^3_{\mathbb{F}_{q^s}}$. Using the same definition of the zeta-function in terms of the N_s as before, compute the zeta-function of the set of lines in projective 3-space.

13. Using Exercise 12 of §IV.4, prove that Dwork's theorem, together with Lemma 1 above, imply that Dwork's theorem holds with "coefficients in $\mathbb{Q}$" replaced by "coefficients in $\mathbb{Z}$ and constant term 1."

14. Let H_f be given by $X_2{}^2 = X_1{}^5 + 1$, and let $p \equiv 3$ or $7 \pmod{10}$. Assuming the Weil Conjectures for the genus 2 curve $\tilde{H}_{\tilde{f}}$, prove that

$$Z(\tilde{H}_{\tilde{f}}/\mathbb{F}_p; T) = \frac{1 + p^2 T^4}{(1 - T)(1 - pT)}.$$

15. Let $\tilde{H}_{\tilde{f}}$ be a smooth projective curve. Assuming the Weil Conjectures for $Z(\tilde{H}_{\tilde{f}}/\mathbb{F}_q; T)$ (which were proved for curves much earlier than for the general case), show that all of the zeros of the complex function of a complex variable

$$F(s) \underset{\text{def}}{=} Z(\tilde{H}_{\tilde{f}}/\mathbb{F}_q; q^{-s})$$

are on the line Re $s = \frac{1}{2}$. This explains the name "Riemann Hypothesis" for part (iii) of the Weil Conjectures.

2. Characters and their lifting

An Ω-*valued character* of a finite group G is a homomorphism from G to the multiplicative group $\Omega^\times$ of nonzero numbers in Ω. Since $g^{\text{order of } g} = 1$ for every $g \in G$, it follows that the image of G under a character must be roots of unity in Ω. For example, if G is the additive group of $\mathbb{F}_p$, if ε is a pth root of 1 in Ω, and if $\tilde{a}$ denotes the least nonnegative residue of $a \in \mathbb{F}_p$, then the map $a \mapsto \varepsilon^{\tilde{a}}$ is a character of $\mathbb{F}_p$. In what follows, we shall omit the tilde and simply write $a \mapsto \varepsilon^a$. If $\varepsilon \neq 1$, then the character is "nontrivial," i.e., its image is not just 1.

If $\mathbb{F}_q$ is a finite field with $q = p^s$ elements, we know that there are $s = [\mathbb{F}_q : \mathbb{F}_p]$ automorphisms $\sigma_0, \ldots, \sigma_{s-1}$ of $\mathbb{F}_q$ given by: $\sigma_i(a) = a^{p^i}$ for $a \in \mathbb{F}_q$ (see Exercise 6 of §III.1). If $a \in \mathbb{F}_q$, by the *trace* of a, written $\text{Tr } a$, we mean

$$\text{Tr } a \underset{\text{def}}{=} \sum_{i=0}^{s-1} \sigma_i(a) = a + a^p + a^{p^2} + \cdots + a^{p^{s-1}}.$$

It is easy to see that $(\text{Tr } a)^p = \text{Tr } a$, i.e., $\text{Tr } a \in \mathbb{F}_p$, and that $\text{Tr}(a + b) = \text{Tr } a + \text{Tr } b$. It then follows that the map

$$a \mapsto \varepsilon^{\text{Tr } a}$$

is an Ω-valued character of the additive group of $\mathbb{F}_q$.

Recall that for every $a \in \mathbb{F}_q$ we have a unique Teichmüller representative $t \in \Omega$, lying in the unramified extension K of $\mathbb{Q}_p$ generated by the $(q - 1)$st roots of 1, such that $t^q = t$ and a is the reduction of t mod p. Our purpose in this section is to find a p-adic power series $\Theta(T)$ whose value at $T = t$ equals $\varepsilon^{\text{Tr } a}$. (More precisely, we'll get the value of $\Theta(T)\Theta(T^p)\Theta(T^{p^2})\cdots$ $\Theta(T^{p^{s-1}})$, where $q = p^s$, to be $\varepsilon^{\text{Tr } a}$ at $T = t$.)

Now fix $a \in \mathbb{F}_q^\times$, and let $t \in K$ be the corresponding Teichmüller representative. Let Tr_K denote the trace over $\mathbb{Q}_p$ of an element of K, i.e., the sum of its conjugates. Then for our Teichmüller representative t we have (see Exercise 1 at the end of §4 below)

$$\text{Tr}_K t = t + t^p + t^{p^2} + \cdots + t^{p^{s-1}} \in \mathbb{Z}_p,$$

and the reduction mod p of $\text{Tr}_K t$ is

$$a + a^p + a^{p^2} + \cdots + a^{p^{s-1}} = \text{Tr } a \in \mathbb{F}_p.$$

Hence, since ε raised to a power in $\mathbb{Z}_p$ depends only on the congruence class mod p, we can write: $\varepsilon^{\text{Tr } a} = \varepsilon^{\text{Tr}_K t}$.

Let $\lambda = \varepsilon - 1$. We have seen that $\text{ord}_p \lambda = 1/(p - 1)$ (see Exercise 7 of §III.4). We want a p-adic expression in t for

$$(1 + \lambda)^{t + t^p + t^{p^2} + \cdots + t^{p^{s-1}}} = \varepsilon^{\text{Tr } a}.$$

The naive thing to do would be to let

$$g(T) = (1 + \lambda)^T = \sum_{i=0}^{\infty} \frac{T(T-1)\cdots(T-i+1)}{i!} \lambda^i,$$

and then take the series $g(T)g(T^p)g(T^{p^2})\cdots g(T^{p^{s-1}})$. But the problem is: how to make sense out of the infinite sum $g(T)$ for the values $T = t$ that interest us. Namely, as soon as $t \notin \mathbb{Z}_p$, i.e., its residue a isn't in $\mathbb{F}_p$, then clearly $|t - i|_p = 1$ for all $i \in \mathbb{Z}$; then

$$\text{ord}_p \frac{t(t-1)\cdots(t-i+1)}{i!} \lambda^i = i\,\text{ord}_p \lambda - \frac{i - S_i}{p-1} = \frac{S_i}{p-1},$$

which does not $\to \infty$.

What we have to do is use the better behaved series $F(X, Y)$ introduced at the end of §IV.2:

$$F(X, Y) = (1 + Y)^X (1 + Y^p)^{(X^p - X)/p} (1 + Y^{p^2})^{(X^{p^2} - X^p)/p^2} \cdots$$
$$\times (1 + Y^{p^n})^{(X^{p^n} - X^{p^{n-1}})/p^n} \cdots,$$

where recall that each term on the right is understood in the sense of the corresponding binomial series in $\mathbb{Q}[[X, Y]]$. We now consider $F(X, Y)$ as a series in X for each fixed Y:

$$F(X, Y) = \sum_{n=0}^{\infty} \left(X^n \sum_{m=n}^{\infty} a_{m,n} Y^m \right), \qquad a_{m,n} \in \mathbb{Z}_p,$$

where we're using the fact that $a_{m,n} \neq 0$ only when $m \geq n$; this follows because each term in the series $B_{(X^{p^n} - X^{p^{n-1}})/p^n, p}(Y^{p^n})$, i.e.,

$$\frac{X^{p^n} - X^{p^{n-1}}}{p^n} \left(\frac{X^{p^n} - X^{p^{n-1}}}{p^n} - 1 \right) \cdots \left(\frac{X^{p^n} - X^{p^{n-1}}}{p^n} - i + 1 \right) \frac{Y^{ip^n}}{i!},$$

has the power of X that appears less than or equal to the power Y^{ip^n} of Y.

Recall that $\lambda = \varepsilon - 1$, and that $\text{ord}_p \lambda = 1/(p-1)$. We set

$$\Theta(T) = F(T, \lambda) = \sum_{n=0}^{\infty} a_n T^n,$$

with $a_n = \sum_{m=n}^{\infty} a_{m,n} \lambda^m$. Clearly $\text{ord}_p a_n \geq n/(p-1)$, since each term in a_n is divisible by λ^n. Also, since the field $\mathbb{Q}_p(\varepsilon) = \mathbb{Q}_p(\lambda)$ is complete, we have $a_n \in \mathbb{Q}_p(\varepsilon)$, and $\Theta(T) \in \mathbb{Q}_p(\varepsilon)[[T]]$. Moreover, $\Theta(T)$ converges for $t \in D(p^{1/(p-1)-})$, because $\text{ord}_p a_n \geq n/(p-1)$.

For our fixed t we now consider the series

$$(1 + Y)^{t + t^p + \cdots + t^{p^{s-1}}} \underset{\text{def}}{=} B_{t + t^p + \cdots + t^{p^{s-1}}, p}(Y).$$

It is easy to prove the following formal identity in $\Omega[[Y]]$:

$$(1 + Y)^{t + t^p + \cdots + t^{p^{s-1}}} = F(t, Y)F(t^p, Y)\cdots F(t^{p^{s-1}}, Y).$$

Namely, after trivial cancellations, the right hand side reduces to

$$(1 + Y)^{t + t^p + \cdots + t^{p^s - 1}}(1 + Y^p)^{(t^{p^s} - t)/p}(1 + Y^{p^2})^{(t^{p^s + 1} - t^p)/p^2}$$

$$\times (1 + Y^{p^3})^{(t^{p^s + 2} - t^{p^2})/p^3} \cdots .$$

Since $t^{p^s} = t$, we get what we want.

Thus, if we substitute t for T in $\Theta(T)\Theta(T^p) \cdots \Theta(T^{p^{s-1}})$, we obtain

$$F(t, \lambda)F(t^p, \lambda) \cdots F(t^{p^{s-1}}, \lambda) = (1 + \lambda)^{t + t^p + \cdots + t^{p^s - 1}}$$

$$= \varepsilon^{\mathrm{Tr}\, a}.$$

To conclude, we have found a nice p-adic power series $\Theta(T) = \sum a_n T^n \in \mathbb{Q}_p(\varepsilon)[[T]]$, satisfying $\mathrm{ord}_p\, a_n \geq n/(p - 1)$, such that the character $a \mapsto \varepsilon^{\mathrm{Tr}\, a}$ of $\mathbb{F}_q$ can be obtained by evaluating $\Theta(T) \cdot \Theta(T^p) \cdots \Theta(T^{p^{s-1}})$ at the Teichmüller lifting of a. Θ can be thought of as "lifting" the character of $\mathbb{F}_q$ to a function on Ω (more precisely, on some disc in Ω, which includes the *closed* unit disc, and hence all Teichmüller representatives).

Liftings such as Θ are important because concepts of analysis often apply directly only to p-adic fields, not to finite fields. If a situation involving finite fields—such as zeta-functions of hypersurfaces defined over finite fields—can be lifted to p-adic fields, we can then do analysis with them. Notice how important it is that our lifting Θ converge at least on the *closed* unit disc (rather than, say, only on the open unit disc): the points we're mainly interested in, the Teichmüller representatives, lie precisely at radius 1.

3. A linear map on the vector space of power series

Let R denote the ring of formal power series over Ω in n indeterminates:

$$R \underset{\mathrm{def}}{=} \Omega[[X_1, X_2, \ldots, X_n]].$$

A monomial $X_1^{u_1} X_2^{u_2} \cdots X_n^{u_n}$ will be denoted X^u, where u is the n-tuple of nonnegative integers $(u_1, \ldots, u_n)$. An element of R is then written $\sum a_u X^u$, where u runs through the set U of all ordered n-tuples of nonnegative integers, and where $a_u \in \Omega$.

Notice that R is an infinite dimensional vector space over Ω. For each $G \in R$ we define a linear map from R to R, also denoted G, by

$$r \mapsto Gr,$$

i.e., multiplying power series in R by the fixed power series G.

Next, for any positive integer q (in our application q will be a power of the prime p), we define a linear map $T_q: R \to R$ by:

$$r = \sum a_u X^u \mapsto T_q(r) = \sum a_{qu} X^u,$$

where qu denotes the n-tuple $(qu_1, qu_2, \ldots, qu_n)$. For example, if $n = 1$, this is the map on power series which forgets about all X^j-terms for which j is not divisible by q and replaces X^j by $X^{j/q}$ in the X^j-terms for which $q|j$.

Now let $\Psi_{q,G} \underset{\text{def}}{=} T_q \circ G : R \to R$. If $G = \sum_{w \in U} g_w X^w$, then $\Psi_{q,G}$ is the linear map defined on elements X^u by

$$\Psi_{q,G}(X^u) = T_q\left(\sum_{w \in U} g_w X^{w+u}\right) = \sum_{v \in U} g_{qv-u} X^v.$$

(Here if the n-tuple $qv - u$ is not in U, i.e., if it has a negative component, we take g_{qv-u} to be zero.)

Let $G_q(X)$ denote the power series $G(X^q) = \sum_{w \in U} g_w X^{qw}$. The following relation is easy to check (Exercise 7 below):

$$G \circ T_q = T_q \circ G_q = \Psi_{q,G_q}.$$

We define the function $| \ |$ on U by: $|u| = \sum_{i=1}^n u_i$. Let

$$R_0 \underset{\text{def}}{=} \left\{ G = \sum_{w \in U} g_w X^w \in R \mid \text{for some } M > 0, \ \text{ord}_p\, g_w \geq M|w| \text{ for all } w \in U \right\}.$$

It is not hard to check that R_0 is closed under multiplication and under the map $G \mapsto G_q$. Note that power series in R_0 must converge when all the variables are in a disc strictly bigger then $D(1)$. An important example of a power series in R_0 is $\Theta(aX^w)$, where X^w is any monomial in $X_1, \ldots, X_n$ and a is in $D(1)$ (see Exercise 2 below).

If V is a finite dimensional vector space over a field F, and if $\{a_{ij}\}$ is the matrix of a linear map $A : V \to V$ with respect to a basis, then the trace of A is defined as

$$\text{Tr } A \underset{\text{def}}{=} \sum a_{ii},$$

i.e., the sum of the elements on the main diagonal (this sum is independent of the choice of basis—for details on this and other basic concepts of linear algebra, see Herstein, *Topics in Algebra*, Ch. 6). (The use of the same symbol Tr as for the trace of an element in $\mathbb{F}_q$ should not cause confusion, since it will always be clear from the context what is meant.) If F has a metric, we can consider the traces of *infinite* matrices A, provided that the corresponding sum $\sum_{i=1}^\infty a_{ii}$ converges.

Lemma 3. *Let $G \in R_0$, and let $\Psi = \Psi_{q,G}$. Then $\text{Tr}(\Psi^s)$ converges for $s = 1, 2, 3, \ldots$, and*

$$(q^s - 1)^n \text{Tr}(\Psi^s) = \sum_{\substack{x \in \Omega^n \\ x^{q^s-1}=1}} G(x) \cdot G(x^q) \cdot G(x^{q^2}) \cdots G(x^{q^{s-1}}),$$

where $x = (x_1, \ldots, x_n)$; $x^{q^i} = (x_1^{q^i}, \ldots, x_n^{q^i})$; and $x^{q^s-1} = 1$ means $x_j^{q^s-1} = 1$ for $j = 1, 2, \ldots, n$.

PROOF. We first prove the lemma for $s = 1$, and then easily reduce the general case to this special case. Since $\Psi(X^u) = \sum_{v \in U} g_{qv-u} X^v$, we have

$$\text{Tr } \Psi = \sum_{u \in U} g_{(q-1)u},$$

which clearly converges by the definition of R_0.

Next, we consider the right-hand side of the equation in the lemma. First of all, for each $i = 1, 2, \ldots, n$, we have

$$\sum_{\substack{x_i \in \Omega \\ x_i^{q-1} = 1}} x_i^{w_i} = \begin{cases} q - 1, & \text{if } q - 1 \text{ divides } w_i \\ 0, & \text{otherwise.} \end{cases}$$

(See Exercise 6 below.) Hence,

$$\sum_{\substack{x \in \Omega^n \\ x_i^{q-1} = 1}} x^w = \prod_i \left(\sum_{x_i^{q-1} = 1} x_i^{w_i} \right) = \begin{cases} (q - 1)^n, & \text{if } q - 1 \text{ divides } w \\ 0, & \text{otherwise.} \end{cases}$$

Thus,

$$\sum_{x_i^{q-1} = 1} G(x) = \sum_{w \in U} g_w \sum_{x_i^{q-1} = 1} x^w = (q - 1)^n \sum_{u \in U} g_{(q-1)u} = (q - 1)^n \operatorname{Tr} \Psi,$$

which proves the lemma for $s = 1$.

Now suppose that $s > 1$. We have:

$$\Psi^s = T_q \circ G \circ T_q \circ G \circ \Psi^{s-2} = T_q \circ T_q \circ G_q \circ G \circ \Psi^{s-2}$$

$$= T_{q^2} \circ G \cdot G_q \circ \Psi^{s-2} = T_{q^2} \circ T_q \circ (G \cdot G_q)_q G \circ \Psi^{s-3}$$

$$= T_{q^3} \circ G \cdot G_q \cdot G_{q^2} \circ \Psi^{s-3} = \cdots = T_{q^s} \circ G \cdot G_q \cdot G_{q^2} \cdots G_{q^{s-1}}$$

$$= \Psi_{q^s, G \cdot G_q \cdot G_{q^2} \cdots G_{q^{s-1}}}.$$

Thus, replacing q by q^s and G by $G \cdot G_q \cdot G_{q^2} \cdots G_{q^{s-1}}$ gives the lemma in the general case. $\qquad\qquad\square$

If A is an $r \times r$ matrix with entries in a field F, and if T is an indeterminate, then $(1 - AT)$ (where 1 is the $r \times r$ identity matrix) is an $r \times r$ matrix with entries in $F[T]$. It plays a role in studying the linear map on F^r defined by A, because for any concrete value $t \in F$ of T, the determinant $\det(1 - At)$ vanishes if and only if there exists a nonzero vector $v \in F^r$ such that $0 = (1 - At)v = v - tAv$, i.e., $Av = (1/t)v$, in other words, if and only if $1/t$ is an eigen-value of A. If $A = \{a_{ij}\}$, we have

$$\det(1 - AT) = \sum_{m=0}^{n} b_m T^m,$$

where

$$b_m = (-1)^m \sum_{\substack{1 \leq u_1 < \cdots < u_m \leq r \\ \sigma \text{ a permutation of the } u\text{'s}}} \operatorname{sgn}(\sigma) a_{u_1, \sigma(u_1)} a_{u_2, \sigma(u_2)} \cdots a_{u_m, \sigma(u_m)}.$$

(Here $\operatorname{sgn}(\sigma)$ equals $+1$ or -1 depending on whether the permutation σ is a product of an even or odd number of transpositions, respectively.)

Now suppose that $A = \{a_{ij}\}_{i,j=1}^{\infty}$ is an infinite "square" matrix, and suppose $F = \Omega$. The expression for $\det(1 - AT)$ still makes sense as a formal power series in $\Omega[[T]]$, as long as the expression for b_m, which now becomes an infinite series (i.e., the condition "$\leq r$" is removed from the u_i), is convergent.

We apply these notions to the case when $A = \{g_{qv-u}\}_{u,v \in U}$ is the "matrix" of $\Psi = T_q \circ G$, where $G \in R_0$, i.e., $\mathrm{ord}_p \, g_w \geq M|w|$. We then have the following estimate for the p-adic ordinal of a term in the expression for b_m:

$$\mathrm{ord}_p[g_{q\sigma(u_1)-u_1} \cdot g_{q\sigma(u_2)-u_2} \cdots g_{q\sigma(u_m)-u_m}]$$
$$\geq M[|q\sigma(u_1) - u_1| + |q\sigma(u_2) - u_2| + \cdots + |q\sigma(u_m) - u_m|]$$
$$\geq M[\textstyle\sum q|\sigma(u_1)| - \sum |u_i|] = M(q-1)\sum |u_i|.$$

(Notice that, if G is a series in n indeterminates, then each u_i is an n-tuple of nonnegative integers: $u_i = (u_{i1}, \ldots, u_{in})$, and $|u_i| = \sum_{j=1}^n u_{ij}$.) This shows that

$$\mathrm{ord}_p \, b_m \to \infty \quad \text{as } m \to \infty$$

and also that

$$\frac{1}{m} \, \mathrm{ord}_p \, b_m \to \infty \quad \text{as } m \to \infty.$$

The latter relationship holds because, if we take into account that there are only finitely many u's with a given $|u|$, we find that the average $|u|$ as we run over a set of *distinct* u_i, i.e., $(1/m)\sum_{i=1}^m |u_i|$, must approach ∞.

This proves that

$$\det(1 - AT) = \sum_{m=0}^{\infty} b_m T^m$$

is well-defined (i.e., the series for each b_m converges), and has an infinite radius of convergence.

We now prove another important auxiliary result, first for finite matrices and then for $\{g_{qv-u}\}$. Namely, we have the following identity of formal power series in $\Omega[[T]]$:

$$\det(1 - AT) = \exp_p\left(-\sum_{s=1}^{\infty} \mathrm{Tr}(A^s) T^s / s\right).$$

To prove this, we first recall from the theory of matrices (Herstein, Ch. 6) that the determinant and trace are unchanged if we conjugate by an invertible matrix: $A \mapsto CAC^{-1}$, i.e., they are invariant under a change of basis. Moreover, over an algebraically closed field such as Ω, a change of basis can be found so that A is upper triangular (for example, in Jordan canonical form), in other words, so that there are no nonzero entries below the main diagonal. So without loss of generality, suppose $A = \{a_{ij}\}_{i,j=1}^r$ is upper triangular. Then the left hand side of the above equality takes the form $\prod_{i=1}^r (1 - a_{ii}T)$. Meanwhile, since $\mathrm{Tr}(A^s) = \sum_{i=1}^r a_{ii}^s$, the right hand side is

$$\exp_p\left(-\sum_{s=1}^{\infty}\sum_{i=1}^r a_{ii}^s T^s/s\right) = \prod_{i=1}^r \exp_p\left(-\sum_{s=1}^{\infty} (a_{ii}T)^s/s\right)$$
$$= \prod_{i=1}^r \exp_p(\log_p(1 - a_{ii}T)) = \prod_{i=1}^r (1 - a_{ii}T).$$

We leave the extension to the case when A is an infinite matrix as an exercise (Exercise 8 below).

We summarize all of this in the following lemma.

Lemma 4. *If* $G(X) = \sum_{w \in U} g_w X^w \in R_0$ *and* $\Psi = T_q \circ G$, *so that* Ψ *has matrix* $A = \{g_{qv-u}\}_{v, u \in U}$, *then the series* $\det(1 - AT)$ *is a well-defined element of* $\Omega[[T]]$ *with infinite radius of convergence, and is equal to*

$$\exp_p\left\{ - \sum_{s=1}^{\infty} \operatorname{Tr}(A^s) T^s / s \right\}.$$

4. p-adic analytic expression for the zeta-function

We now prove that for any hypersurface H_f defined by $f(X_1, \ldots, X_n) \in \mathbb{F}_q[X_1, \ldots, X_n]$, the zeta-function

$$Z(H_f/\mathbb{F}_q; T) \in \mathbb{Z}[[T]] \subset \Omega[[T]]$$

is a quotient of two power series in $\Omega[[T]]$ with infinite radius of convergence. (Alternate terminology: is p-adic *meromorphic*, is a quotient of two p-adic *entire* functions.)

We prove this by induction on the number n of variables (i.e., on the dimension $n - 1$ of the hypersurface H_f). The assertion is trivial if $n = 0$ (i.e., H_f is the empty set). Suppose it holds for $1, 2, \ldots, n - 1$ variables. We claim that, instead of proving our assertion for

$$Z(H_f/\mathbb{F}_q; T) = \exp\left(\sum N_s T^s / s\right),$$

it suffices to prove it for

$$Z'(H_f/\mathbb{F}_q; T) \underset{\text{def}}{=} \exp\left(\sum_{s=1}^{\infty} N_s' T^s / s\right),$$

where

$N_s' \underset{\text{def}}{=}$ number of $(x_1, \ldots, x_n) \in \mathbb{F}_{q^s}^n$ such that $f(x_1, \ldots, x_n) = 0$

and all of the x_i are nonzero

$=$ number of $(x_1, \ldots, x_n) \in \mathbb{F}_{q^s}^n$ such that $f(x_1, \ldots, x_n) = 0$

and $x_i^{q^s - 1} = 1$, $i = 1, \ldots, n$.

How does $Z'(H_f/\mathbb{F}_q; T)$ differ from $Z(H_f/\mathbb{F}_q; T)$? Well,

$$Z(H_f/\mathbb{F}_q; T) = Z'(H_f/\mathbb{F}_q; T) \cdot \exp\left(\sum (N_s - N_s') T^s / s\right),$$

and the exp factor on the right is the zeta-function for the union of the n hypersurfaces H_i ($i = 1, \ldots, n$) defined by $f(X_1, \ldots, X_n) = 0$ and $X_i = 0$. Note that H_i either is a copy of $(n - 1)$-dimensional affine space given in

$\mathbb{A}^n_{\mathbb{F}_q}$ by the equation $x_i = 0$ (this is the case if $f(X_1, \ldots, X_n)$ is divisible by X_i), or else is a lower dimensional $((n-2)$-dimensional$)$ hypersurface. In the first case we know its zeta-function explicitly (see Exercise 1 of §1), and in the latter case we know that its zeta-function is meromorphic by the induction assumption. Now the zeta-function for the union of the H_i is easily seen to be the product of the individual zeta-functions of the H_i, *divided* by the product of the zeta-functions of the overlaps of H_i and H_j ($i \neq j$)—i.e., the hypersurface in a copy of $\mathbb{A}^{n-2}_{\mathbb{F}_q}$ defined by $X_i = X_j = 0$ and $f(X_1, \ldots, X_n) = 0$—*multiplied* by the product of the zeta-functions of the triple overlaps, *divided* by the product of the zeta-functions of the quadruple overlaps, and so on. But all of these zeta-functions are *p*-adic meromorphic by the induction assumption or by the explicit formula for the zeta-function of affine space. Hence, if Z' is proved to be *p*-adic meromorphic, it then follows that Z is *p*-adic meromorphic as well. See Exercises 4–5 of §1 for a similar argument.

Fix an integer $s \geq 1$. Let $q = p^r$. Recall that, if t denotes the Teichmüller representative of $a \in \mathbb{F}_{q^s}$, then the *p*th root of 1 given by $\varepsilon^{\mathrm{Tr}\, a}$ has a *p*-adic analytic formula in terms of t:

$$\varepsilon^{\mathrm{Tr}\, a} = \Theta(t)\Theta(t^p)\Theta(t^{p^2}) \cdots \Theta(t^{p^{rs-1}}).$$

A basic and easily proved fact about characters (see Exercises 3–5 below) is that

$$\sum_{x_0 \in \mathbb{F}_{q^s}} \varepsilon^{\mathrm{Tr}(x_0 u)} = \begin{cases} 0, & \text{if } u \in \mathbb{F}_{q^s}^\times \\ q^s, & \text{if } u = 0; \end{cases}$$

and so, if we subtract the $x_0 = 0$ term,

$$\sum_{x_0 \in \mathbb{F}_{q^s}^\times} \varepsilon^{\mathrm{Tr}(x_0 u)} = \begin{cases} -1, & \text{if } u \in \mathbb{F}_{q^s}^\times \\ q^s - 1, & \text{if } u = 0. \end{cases}$$

Applying this to $u = f(x_1, \ldots, x_n)$ and summing over all $x_1, \ldots, x_n \in \mathbb{F}_{q^s}^\times$, we obtain

$$\sum_{x_0, x_1, \ldots, x_n \in \mathbb{F}_{q^s}^\times} \varepsilon^{\mathrm{Tr}(x_0 f(x_1, \ldots, x_n))} = q^s N_s' - (q^s - 1)^n.$$

Now replace all of the coefficients in $X_0 f(X_1, \ldots, X_n) \in \mathbb{F}_q[X_0, X_1, \ldots, X_n]$ by their Teichmüller representatives to get a series $F(X_0, X_1, \ldots, X_n) = \sum_{i=1}^N a_i X^{w_i} \in \Omega[X_0, X_1, \ldots, X_n]$, where X^{w_i} denotes $X_0^{w_{i0}} X_1^{w_{i1}} \cdots X_n^{w_{in}}$, $w_i = (w_{i0}, w_{i1}, \ldots, w_{in})$.

We obtain:

$$q^s N_s' = (q^s - 1)^n + \sum_{x_0, x_1, \ldots x_n \in \mathbb{F}_{q^s}^\times} \varepsilon^{\mathrm{Tr}(x_0 f(x_1, \ldots, x_n))}$$

$$= (q^s - 1)^n + \sum_{\substack{x_0, x_1, \ldots, x_n \in \Omega \\ x_0^{q^s-1} = \cdots = x_n^{q^s-1} = 1}} \prod_{i=1}^N \Theta(a_i x^{w_i})\Theta(a_i^p x^{p w_i}) \cdots$$

$$\cdot \, \Theta(a_i^{p^{rs-1}} x^{p^{rs-1} w_i}).$$

Note that, since $f(X_1, \ldots, X_n)$ has coefficients in $\mathbb{F}_q$, $q = p^r$, it follows that $a_i^{p^r} = a_i$. Now let

$$G(X_0, \ldots, X_n) \underset{\text{def}}{=} \prod_{i=1}^{N} \Theta(a_i X^{w_i}) \Theta(a_i^p X^{pw_i}) \cdots \Theta(a_i^{p^{r-1}} X^{p^{r-1}w_i}),$$

so that

$$q^s N_s' = (q^s - 1)^n + \sum_{\substack{x_0, x_1, \ldots, x_n \in \Omega \\ x_0^{q^s-1} = \cdots = x_n^{q^s-1} = 1}} G(x) \cdot G(x^q) \cdot G(x^{q^2}) \cdots G(x^{q^{s-1}}).$$

Since the series $\Theta(a_i^{p^l} X^{p^l w_i})$ are each in R_0 (see Exercise 2 below), so is G:

$$G(X_0, \ldots, X_1) \in R_0 \subset \Omega[[X_0, \ldots, X_n]].$$

Hence, by Lemma 3, we have

$$q^s N_s' = (q^s - 1)^n + (q^s - 1)^{n+1} \operatorname{Tr}(\Psi^s),$$

i.e.,

$$N_s' = \sum_{i=0}^{n} (-1)^i \binom{n}{i} q^{s(n-i-1)} + \sum_{i=0}^{n+1} (-1)^i \binom{n+1}{i} q^{s(n-i)} \operatorname{Tr}(\Psi^s).$$

If we set (recall: A is the matrix of Ψ)

$$\Delta(T) \underset{\text{def}}{=} \det(1 - AT) = \exp_p\left\{-\sum_{s=1}^{\infty} \operatorname{Tr}(\Psi^s) T^s/s,\right\},$$

we conclude that

$$Z'(H_f/\mathbb{F}_q; T) = \exp_p\left\{\sum_{s=1}^{\infty} N_s' T^s/s\right\}$$

$$= \prod_{i=0}^{n} \left[\exp_p\left\{\sum_{s=1}^{\infty} q^{s(n-i-1)} T^s/s\right\}\right]^{(-1)^i\binom{n}{i}} \cdot$$

$$\times \prod_{i=0}^{n+1} \left[\exp_p\left\{\sum_{s=1}^{\infty} q^{s(n-i)} \operatorname{Tr}(\Psi^s) T^s/s\right\}\right]^{(-1)^i\binom{n+1}{i}}$$

$$= \prod_{i=0}^{n} (1 - q^{n-i-1} T)^{(-1)^{i+1}\binom{n}{i}} \prod_{i=0}^{n+1} \Delta(q^{n-i} T)^{(-1)^{i+1}\binom{n+1}{i}}.$$

By Lemma 4, each term in this "alternating product" is a p-adic entire function.

This concludes the proof that the zeta-function is p-adic meromorphic. This result is the heart of the proof of Dwork's theorem. In the next section we finish the proof that the zeta-function is a quotient of two polynomials.

EXERCISES

1. Let $t \in \Omega$ be a primitive $(p^s - 1)$th root of 1. Prove that the conjugates of t over $\mathbb{Q}_p$ are precisely: $t, t^p, t^{p^2}, \ldots, t^{p^{s-1}}$. In other words, the conjugates of the Teichmüller representative of $a \in \mathbb{F}_q$ are the Teichmüller representatives of the conjugates of a over $\mathbb{F}_p$.

2. Let $X^w = X_1^{w_1} \cdots X_n^{w_n}$, and let $a \in D(1)$. Prove that $\Theta(aX^w) \in R_0$.

3. Let $\sigma_1, \ldots, \sigma_s$ be distinct automorphisms of a field K. Prove that there is no nonzero linear combination $\sum a_i\sigma_i$ such that $\sum a_i\sigma_i(x) = 0$ for every $x \in K$.

4. Let $\varepsilon \in \Omega$ be a primitive pth root of 1. Prove that $\sum_{x \in \mathbb{F}_q} \varepsilon^{\mathrm{Tr}\, x} = 0$.

5. Prove that
$$\sum_{x \in \mathbb{F}_{q^s}^x} \varepsilon^{\mathrm{Tr}(xu)} = \begin{cases} -1, & \text{if } u \in \mathbb{F}_{q^s}^x, \\ q^s - 1, & \text{if } u = 0. \end{cases}$$

6. Prove that for any positive integers n and a,
$$\sum_{\zeta \in \Omega,\, \zeta^n = 1} \zeta^a = \begin{cases} n, & \text{if } n \text{ divides } a; \\ 0, & \text{otherwise.} \end{cases}$$

7. Prove that $G \circ T_q = T_q \circ G_q$ in the notation of §V.3.

8. Extend the result
$$\det(1 - AT) = \exp_p\left(-\sum_{s=1}^{\infty} \mathrm{Tr}(A^s)T^s/s\right)$$
to infinite matrices A, with a suitable hypothesis on convergence of Tr.

9. *A review problem.* Let $f(X) = \sum_{i=0}^{n} a_i X^i \in \mathbb{F}_q[X]$ be a polynomial in one variable with coefficients in $\mathbb{F}_q$, $q = p^r$, and nonzero constant term. We are interested in the number N of distinct roots of $f(X)$ in $\mathbb{F}_q$. For each $i = 0$, $1, \ldots, n$, let A_i be the Teichmüller representative of a_i. Let $\varepsilon = 1 + \lambda$ be a fixed primitive pth root of 1 in Ω, and define $\Theta(T)$ as in §V.2. Let
$$G(X, Y) = \prod_{i=0}^{n} \prod_{j=0}^{r-1} \Theta(A_i^{p^j} X^{ip^j} Y^{p^j}).$$

Prove that
$$N = \frac{q-1}{q} + \frac{1}{q} \sum_{\substack{x,\, y \in \Omega \\ x^{q-1} = y^{q-1} = 1}} G(x, y).$$

5. The end of the proof

Dwork's theorem will now follow easily once we prove the following criterion for a power series to be a rational function.

Lemma 5. *Let* $F(T) = \sum_{i=0}^{\infty} a_i T^i \in K[[T]]$, *where* K *is any field. For* $m, s \geq 0$, *let* $A_{s,m}$ *be the matrix* $\{a_{s+i+j}\}_{0 \leq i,j \leq m}$:
$$\begin{pmatrix} a_s & a_{s+1} & a_{s+2} & \cdots & a_{s+m} \\ a_{s+1} & a_{s+2} & a_{s+3} & \cdots & a_{s+m+1} \\ a_{s+2} & a_{s+3} & a_{s+4} & \cdots & a_{s+m+2} \\ \vdots & \vdots & \vdots & & \vdots \\ a_{s+m} & a_{s+m+1} & a_{s+m+2} & \cdots & a_{s+2m} \end{pmatrix},$$
and let $N_{s,m} \underset{\mathrm{def}}{=} \det(A_{s,m})$. *Then* $F(T)$ *is a quotient of two polynomials*
$$F(T) = \frac{P(T)}{Q(T)}, \qquad P(T), Q(T) \in K[T],$$

if and only if there exist integers $m \geq 0$ and S such that $N_{s,m} = 0$ whenever $s \geq S$.

PROOF. First suppose that $F(T)$ is such a quotient. Let $P(T) = \sum_{i=0}^{M} b_i T^i$, $Q(T) = \sum_{i=0}^{N} c_i T^i$. Then, since $F(T) \cdot Q(T) = P(T)$, equating coefficients of T^i for $i > \max(M, N)$ gives:

$$\sum_{j=0}^{N} a_{i-N+j} c_{N-j} = 0.$$

Let $S = \max(M - N + 1, 1)$, and let $m = N$. If $s \geq S$, we write this equation for $i = s + N, s + N + 1, \ldots, s + 2N$:

$$a_s c_N + a_{s+1} c_{N-1} + \cdots + a_{s+N} c_0 = 0$$
$$a_{s+1} c_N + a_{s+2} c_{N-1} + \cdots + a_{s+N+1} c_0 = 0$$
$$\vdots$$
$$a_{s+N} c_N + a_{s+N+1} c_{N-1} + \cdots + a_{s+2N} c_0 = 0.$$

Hence the matrix of coefficients of the c_j, which is $A_{s,N}$, has zero determinant:

$$N_{s,m} = N_{s,N} = 0 \quad \text{for } s \geq S.$$

Conversely, suppose that $N_{s,m} = 0$ for $s \geq S$, where m is chosen to be minimal with this property that $N_{s,m} = 0$ for all s larger than some S. We claim that $N_{s,m-1} \neq 0$ for *all* $s \geq S$.

Suppose the contrary. Then some linear combination of the first m rows $r_0, r_1, \ldots, r_{m-1}$ of $A_{s,m}$ vanish in all but perhaps the last column. Let r_{i_0} be the first row having nonzero coefficient in this linear combination, i.e., the i_0th row r_{i_0} can be expressed as

$$\alpha_1 r_{i_0+1} + \alpha_2 r_{i_0+2} + \cdots + \alpha_{m-i_0-1} r_{m-1}$$

except perhaps in the last column. In $A_{s,m}$ replace r_{i_0} by $r_{i_0} - (\alpha_1 r_{i_0+1} + \cdots + \alpha_{m-i_0-1} r_{m-1})$, and consider two cases:

(1) $i_0 > 0$. Then we have a matrix of the form

$$\begin{pmatrix} a_s & a_{s+1} & \cdots & & a_{s+m} \\ a_{s+1} & a_{s+2} & \cdots & & a_{s+m+1} \\ \vdots & \vdots & & & \vdots \\ 0 & 0 & \cdots & 0 & \beta \\ \vdots & \vdots & & & \vdots \\ a_{s+m} & a_{s+m+1} & \cdots & & a_{s+2m} \end{pmatrix},$$

and the boxed matrix has determinant $N_{s+1,m-1} = 0$.

(2) $i_0 = 0$. Then we have:

$$\begin{pmatrix} 0 & 0 & \cdots & 0 & \beta \\ a_{s+1} & a_{s+2} & \cdots & & a_{s+m+1} \\ \vdots & \vdots & & & \vdots \\ a_{s+m} & a_{s+m+1} & \cdots & & a_{s+2m} \end{pmatrix},$$

Now $N_{s+1,m-1}$ is the determinant of either of the boxed matrices. Since the determinant $N_{s,m}$ of the entire matrix is 0, either the determinant of the lower left boxed matrix is 0, or else $\beta = 0$, in which case also $N_{s+1,m-1} = 0$.

Thus, in either case $N_{s+1,m-1} = 0$, and, by induction, we may obtain $N_{s',m-1} = 0$ for all $s' \geq s$. This contradicts our choice of m to be minimal.

But then for any $s \geq S$ we have $N_{s,m} = 0$ and $N_{s,m-1} \neq 0$. Hence there is a linear combination of the rows of $A_{s,m}$ which vanishes, *in which the coefficient of the last row is nonzero*. Thus, the last row of $A_{s,m}$ for any $s \geq S$ is a linear combination of the preceding m rows. Hence any simultaneous solution to

$$
\begin{array}{llll}
a_s u_m & + \; a_{s+1} u_{m-1} & + \cdots + \; a_{s+m} u_0 & = 0 \\
\;\;\vdots & \quad\;\; \vdots & \qquad\;\; \vdots & \\
a_{s+m-1} u_m & + \; a_{s+m} u_{m-1} & + \cdots + \; a_{s+2m-1} u_0 & = 0
\end{array}
$$

is also a solution to

$$
a_{s+m} u_m + a_{s+m+1} u_{m-1} + \cdots + a_{s+2m} u_0 = 0,
$$

and, by induction, to every

$$
a_s u_m + a_{s+1} u_{m-1} + \cdots + a_{s+m} u_0 = 0
$$

for $s \geq S$. This clearly implies that

$$
\left(\sum_{i=0}^{m} u_i X^i \right) \cdot \left(\sum_{i=1}^{\infty} a_i X^i \right)
$$

is a polynomial (of degree $< S + m$).   □

We now use Lemma 5 to prove the theorem. We must make use of the "p-adic Weierstrass Preparation Theorem" (Theorem 14, §IV.4). In the form we need, it says that, if $F(T)$ is a p-adic entire function, then for any R there exists a polynomial $P(T)$ and a p-adic power series $F_0(T) \in 1 + T\Omega[[T]]$ which converges, *along with its reciprocal* $G(T)$, on the disc $D(R)$ of radius R, such that $F(T) = P(T) \cdot F_0(T)$. Namely, in Theorem 14 let $p^\lambda = R$; since F is entire, it converges on $D(p^\lambda)$.

For brevity, let $Z(T) = Z(H_i/\mathbb{F}_q; T)$. We know from §4 that we can write $Z(T) = A(T)/B(T)$, where $A(T)$ and $B(T)$ are p-adic entire functions. Choose $R > q^n$; for simplicity, take $R = q^{2n}$. If we apply the fact in the last paragraph to $B(T)$, we may write $B(T) = P(T)/G(T)$, where $G(T)$ converges on $D(R)$. Let $F(T) = A(T) \cdot G(T)$, which converges on $D(R)$. Thus,

$$
F(T) = P(T) \cdot Z(T).
$$

Let $F(T) = \sum_{i=0}^{\infty} b_i T^i \in 1 + T\Omega[[T]]$, $P(T) = \sum_{i=0}^{v} c_i T^i \in 1 + T\Omega[T]$, $Z(T) = \sum_{i=0}^{\infty} a_i T^i \in 1 + T\mathbb{Z}[[T]]$. By Lemma 2 of §1, we have

$$
|a_i|_{\infty} \leq q^{in}.
$$

Since $F(T)$ converges on $D(R)$, we also have for i sufficiently large:

$$|b_i|_p \leq R^{-i} = q^{-2ni}.$$

Choose $m > 2e$. Then fix m. Let $A_{s,m} = \{a_{s+i+j}\}_{0 \leq i,j \leq m}$, as before, and $N_{s,m} = \det(A_{s,m})$. We claim that for our m we have $N_{s,m} = 0$ for s sufficiently large. By Lemma 5, this claim will imply that $Z(T)$ is a rational function. Equating coefficients in $F(T) = P(T)Z(T)$, we have:

$$b_{j+e} = a_{j+e} + c_1 a_{j+e-1} + c_2 a_{j+e-2} + \cdots + c_e a_j.$$

In the matrix $A_{s,m}$, we add to each $(j + e)$th column—starting from the last and moving left until the eth column—the linear combination of the previous e columns with coefficient c_k for the $(j + e - k)$th column. This gives us a matrix whose first e columns are the same as in $A_{s,m}$, the rest of its columns have a's replaced by the corresponding b's, and which still has determinant $N_{s,m}$. We use this form of the matrix to estimate $|N_{s,m}|_p$.

Since $a_i \in \mathbb{Z}$, we have $|a_i|_p \leq 1$. Thus, $|N_{s,m}|_p \leq (\max_{j \geq s+e}|b_j|_p)^{m+1-e} < R^{-s(m+1-e)}$ for s sufficiently large. Since $R = q^{2n}$, and $m > 2e$, this gives us: $|N_{s,m}|_p < q^{-ns(m+2)}$.

On the other hand, a crude estimate based directly on $A_{s,m}$ gives: $|N_{s,m}|_\infty \leq (m + 1)! q^{n(s+2m)(m+1)} = (m+1)! q^{2nm(m+1)} q^{ns(m+1)}$. Multiplying together these two estimates, we see that the product of the p-adic norm and the usual absolute value of $N_{s,m}$ is bounded by an expression which is less than 1 for s sufficiently large:

$$|N_{s,m}|_p \cdot |N_{s,m}|_\infty < q^{-ns(m+2)} \cdot (m+1)! q^{2nm(m+1)} q^{ns(m+1)} = \frac{(m+1)! q^{2nm(m+1)}}{q^{ns}} < 1$$

for s sufficiently large. But $N_{s,m} \in \mathbb{Z}$, and the only integer n with the property that $|n| \cdot |n|_p < 1$ is $n = 0$. Hence $N_{s,m} = 0$ for s sufficiently large.

Therefore, $Z(T)$ is a rational function, and Dwork's theorem is proved.

$\square$

Bibliography

Within each section, the order is approximately by increasing difficulty. In the case of books and long articles, this is very rough, and is based on the level of background required to understand the sections most relevant to the material in our Chapters I–V.

(a) *Background*

1. G. Simmons, *Introduction to Topology and Modern Analysis*, McGraw-Hill, 1963.

2. I. Herstein, *Topics in Algebra*, John Wiley and Sons, 1975.

3. S. Lang, *Algebra*, Addison-Wesley, 1965.

4. W. Rudin, *Principles of Mathematical Analysis*, McGraw-Hill, 1976.

(b) *General*

1. Z. I. Borevich and I. R. Shafarevich, *Number Theory* (translated from Russian), Academic Press, 1966.

2. S. Lang, *Algebraic Number Theory*, Addison-Wesley, 1970.

3. J.-P. Serre, *A Course in Arithmetic* (translated from French), Springer-Verlag, 1973.

4. K. Ireland and M. Rosen, *A Classical Introduction to Modern Number Theory*, Springer-Verlag, 1982.

5. E. B. Dynkin and V. A. Uspenskii, *Problems in the Theory of Numbers*, Part Two of *Mathematical Conversations* (translated from Russian), D. C. Heath and Co., 1963.

6. L. Washington, *Introduction to Cyclotomic Fields*, Springer-Verlag, 1982.

7. N. Koblitz, *p-adic Analysis: a Short Course on Recent Work*, Cambridge University Press, 1980.

8. K. Mahler, *Introduction to p-adic Numbers and Their Functions*, Cambridge University Press, 1973.

9. G. Bachman, *Introduction to p-adic Numbers and Valuation Theory*, Academic Press, 1964.

10. A. F. Monna, *Analyse non-archimédienne*, Springer-Verlag, 1970.

11. S. Lang, *Cyclotomic Fields*, Vols. 1 and 2, Springer-Verlag, 1978 and 1980.

(c) *Chapter II*

1. K. Iwasawa, *Lectures on p-adic L-Functions*, Princeton University Press, 1972.

2. T. Kubota and H. W. Leopoldt, Eine *p*-adische Theorie der Zetawerte I, *J. Reine Angew. Math.*, 214/215 (1964), 328–339.

3. N. Katz, *p*-adic *L*-functions via moduli of elliptic curves, *Proceedings A.M.S. Summer Institute of Alg. Geom. at Arcata, Calif.*, 1974.

4. S. Lang, *Introduction to Modular Forms*, Springer-Verlag, 1976.

5. J.-P. Serre, Formes modulaires et fonctions zêta *p*-adiques, *Modular Functions of One Variable III* (Lecture Notes in Math. 350), Springer-Verlag, 1973.

6. Ju. I. Manin, Periods of cusp forms and *p*-adic Hecke series, translated in *Math. USSR-Sb.*, **21** (1973), 371–393. (Note especially §8.)

7. M. M. Višik, Non-Archimedean measures connected with Dirichlet series, translated in *Math. USSR-Sb.*, **28** (1976).

8. Y. Amice and J. Vélu, Distributions *p*-adiques associées aux séries de Hecke, *Journées arithmétiques*, 1974.

9. N. Katz, *p*-adic properties of modular schemes and modular forms, *Modular Functions of One Variable III* (Lecture Notes in Math. 350), Springer-Verlag, 1973.

10. N. Katz, The Eisenstein measure and *p*-adic interpolation, *Amer. J. Math.*, **99** (1977), 238–311.

11. N. Katz, *p*-adic interpolation of real analytic Eisenstein series, *Ann. of Math.*, **104** (1976), 459–571.

12. B. Mazur and P. Swinnerton-Dyer, Arithmetic of Weil curves, *Invent. Math.*, **25** (1974), 1–61.

(d) *Chapter IV*

1. G. Overholtzer, Sum functions in elementary *p*-adic analysis, *Amer. J. Math.*, **74** (1952), 332–346.

2. Y. Morita, A *p*-adic analogue of the Γ-function, *J. Fac. Sci. Univ. Tokyo*, **22** (1975), 255–266.

3. J. Diamond, The *p*-adic log gamma function and *p*-adic Euler constants, *Trans. A.M.S.*, **233** (1977), 321–337.

4. B. Dwork, §1 of: On the zeta function of a hypersurface, *Publ. Math. I.H.E.S.*, **12** (1962), 7–17.

5. Y. Amice, *Les nombres p-adiques*, Presses Universitaires de France, 1975.

6. B. H. Gross and N. Koblitz, Gauss sums and the *p*-adic Γ-function, *Annals of Math.*, **109** (1979), 569–581.

7. B. Ferrero and R. Greenberg, On the behavior of *p*-adic *L*-functions at $s = 0$, *Inventiones Math.*, **50** (1978), 91–102.

(e) *Chapter V*

1. A. Weil, Number of solutions of equations in finite fields, *Bull. Amer. Math. Soc.*, **55** (1949), 497–508.

2. J.-P. Serre, Rationalité des fonctions ζ des variétés algébriques (d'après Bernard Dwork), *Séminaire Bourbaki*, No. 198, February 1960.

3. B. Dwork, On the rationality of the zeta function of an algebraic variety, *Amer. J. Math.*, **82** (1960), 631–648.

4. P. Monsky, *p-adic Analysis and Zeta Functions*, Lectures at Kyoto University, Kinokuniya Book Store, Tokyo, or Brandeis Univ. Math. Dept., 1970.

5. N. Katz, Une formule de congruence pour la fonction ζ, *S.G.A. 7 II* (Lecture Notes in Math. 340), Springer-Verlag, 1973.

6. B. Dwork, On the zeta function of a hypersurface, *Publ. Math. I.H.E.S.*, **12** (1962), 5–68.

7. B. Dwork, On the zeta function of a hypersurface II, *Ann. of Math.*, **80** (1964), 227–299.

8. B. Dwork, A deformation theory for the zeta function of a hypersurface, *Proc. Int. Cong. Math. 1962 Stockholm*, 247–259.

9. N. Katz, Travaux de Dwork, *Séminaire Bourbaki*, No. 409 (Lecture Notes in Math. 317), Springer-Verlag, 1973.

Answers and hints for the exercises

3. Write $\|x + y\|^N = \|(x + y)^N\|$, use the binomial expansion and Property (3) of a norm to get an inequality for $\|x + y\|^N$ in terms of $\max(\|x\|, \|y\|)$, then take Nth roots and let $N \to \infty$.

4. If $x \in F$ has the property that $\|x\| < 1$ and $\|x - 1\| < 1$, and if $\| \ \|$ is non-Archimedean, then $1 = \|1 - x + x\| \le \max(\|x - 1\|, \|x\|) < 1$, a contradiction. Conversely, suppose that $\| \ \|$ is Archimedean. Then by definition, there exist $x, y \in F$ such that $\|x + y\| > \max(\|x\|, \|y\|)$. Let $\alpha = x/(x + y)$, and show that $\|\alpha\| < 1$ and $\|\alpha - 1\| < 1$.

5. Suppose $\| \ \|_1 \sim \| \ \|_2$. Let $a \in F$ be any nonzero element with $\|a\|_2 \ne 1$, say $\|a\|_2 > 1$. Then there is a unique α such that $\|a\|_1 = \|a\|_2^\alpha$. *Claim*: $\|x\|_1 = \|x\|_2^\alpha$ for all $x \in F$. If, say, there were an x with $\|x\|_1 > \|x\|_2^\alpha$ (also suppose that $\|x\|_1 > 1$), then choose large powers x^m and a^n such that $\|x^m/a^n\|_1$ approaches 1; but then show that $\|x^m/a^n\|_2$ approaches 0, and hence the two norms are not equivalent. Finally, note that $\alpha > 0$, or else we would not have $\| \ \|_1 \sim \| \ \|_2$. (The converse direction in this exercise is easy.)

6. If $\rho = 1$, you get the trivial norm. If $\rho > 1$, you don't get a norm at all; for example, choose N large enough so that $\rho^N > 2$ and take $x = 1, y = p^N - 1$. Then check that $\rho^{\text{ord}_p(x+y)} > \rho^{\text{ord}_p x} + \rho^{\text{ord}_p y}$.

7. The sequence $\{p_1^n\}$ approaches 0 in $|\ |_{p_1}$ but not in $|\ |_{p_2}$.

8. The hardest part is to prove the triangle inequality for $|\ |^\alpha, \alpha \le 1$. By supposing $|x| \ge |y|$ and setting $u = y/x$, reduce to showing that

$$-1 \le u \le 1 \quad \text{implies} \quad |1 + u|^\alpha \le 1 + |u|^\alpha,$$

which is true if

$$0 \le u \le 1 \quad \text{implies} \quad f(u) = 1 + u^\alpha - (1 + u)^\alpha \ge 0.$$

Since $f(0) = 0$ and $f(1) \ge 0$, this follows by showing that $f'(u) \ne 0$ on $(0, 1)$.

9. Use Exercises 2 and 3: if $\|n\|_{\text{Arch}} > 1$, then the sequence $\{1/n^i\}$ approaches zero in $\|\ \|_{\text{Arch}}$ but not in $\|\ \|_{\text{non-Arch}}$.

10. Show that the least positive integer of the form $nN + mM$ must be a common divisor of N and M.

11. $3, 7, 1, 1, 7, -2, 0, 0, 3, 2, -2, 0, -1, -1, 4$.

12–14. Prove the lemma: $\text{ord}_p(n!) = \sum[n/p^j]$, where $[\ ]$ is the greatest integer function and the sum is over $j \geq 1$ (note that this is only a finite sum).

15. $1/25, 25, 1, 1/25, 1/243, 1/243, 243, 1/13, 1/7, 1/2, 182, 1/81, 3, 2^{N-2^N}, 1/2$.

16. p does not divide its denominator when x is written as a fraction in lowest terms.

17. Use Exercise 14.

19. Use the "diagonal process," as follows. Choose an infinite subsequence of integers with the same first digit, an infinite subsequence of that with the same first two digits, and so on. Then take the subsequence consisting of the first element from the first subsequence, the second element from the second subsequence, ..., the ith element from the ith subsequence, ...

CHAPTER I §5

1. $(p - a_{-m})p^{-m} + (p - 1 - a_{-m+1})p^{-m+1} + \cdots$
$\quad + (p - 1 - a_0) + (p - 1 - a_1)p + \cdots$.

2.   (i) $4 + 0 \cdot 7 + 1 \cdot 7^2 + 5 \cdot 7^3$
  (ii) $2 + 0 \cdot 5 + 1 \cdot 5^2 + 3 \cdot 5^3$
  (iii) $8 \cdot 11^{-1} + 8 + 9 \cdot 11 + 5 \cdot 11^2$
  (iv) $1 \cdot 2 + \overline{1 \cdot 2^2} + 0 \cdot 2^3$ (the bar denotes repeating digits)
  (v) $1 + 1 \cdot 7 + \overline{1 \cdot 7^2}$
  (vi) $10 + 9 \cdot 11 + \overline{9 \cdot 11^2}$
  (vii) $\overline{10 + 0 \cdot 13 + 4 \cdot 13^2 + 7 \cdot 13^3}$
 (viii) $2 \cdot 5^{-3} + \overline{4 \cdot 5^{-2} + 1 \cdot 5^{-1}}$
  (ix) $2 \cdot 3^2 + 2 \cdot 3^3 + 2 \cdot 3^4 + 2 \cdot 3^5$
  (x) $2 \cdot 3^{-1} + 1 + \overline{1 \cdot 3}$
  (xi) $1 \cdot 2^{-3} + \overline{1 \cdot 2^{-2} + 0 \cdot 2^{-1}}$
  (xii) $4 \cdot 5^{-1} + \overline{4 + 3 \cdot 5}$

4. To prove that $a/b \in \mathbb{Q}$ has repeating digits in its p-adic expansion, first reduce to the case $p \nmid b$. Then first suppose that a/b is between 0 and -1. Multiply the denominator and numerator of a/b by some c which gets the denominator in the form $cb = p^r - 1$ for some r. Let $d = -ac$, so that $0 < d < p^r - 1$. Then $a/b = d/(1 - p^r)$; now expand as a geometric series. You find that a/b has a "purely" repeating expansion. If a/b is not between 0 and -1, then it is obtained by adding or subtracting a positive integer from a purely repeating expansion, and the result will still be a repeating expansion once you're past the first few digits. An alternate proof is to show that in long division you must eventually get repetition in the remainders.

5. The cardinality of the continuum. You can construct a one-to-one correspondence f between $\mathbb{Z}_p$ and the real numbers in the interval $[0, 1]$ written to the base p by setting $f(a_0 + a_1 p + \cdots + a_n p^n + \cdots) = a_0/p + a_1/p^2 + \cdots + a_n/p^{n+1} + \cdots$. ($f$ is not quite one-to-one, since a real number in $(0, 1)$ with a terminating expansion has two preimages; for example, $f(1) = f(-p) = 1/p$.)

7. $1 + 0 \cdot 2 + 1 \cdot 2^2 + 0 \cdot 2^3 + 1 \cdot 2^4 + \cdots$.

8. (i), (iii), (iv), (v), (ix).

9. $2 + 1 \cdot 5 + 2 \cdot 5^2 + 1 \cdot 5^3 + \cdots, \ 3 + 3 \cdot 5 + 2 \cdot 5^2 + 3 \cdot 5^3 + \cdots; \ 2 + 5 \cdot 7 + 0 \cdot 7^2$ $+ 6 \cdot 7^3 + \cdots, \ 5 + 1 \cdot 7 + 6 \cdot 7^2 + 0 \cdot 7^3 + \cdots$.

10. 5, 13, 17.

11. Let $\alpha_1 \in \mathbb{Z}_p^\times$ be any number which is a square mod p, let $\alpha_2 \in \mathbb{Z}_p^\times$ be any number which is not a square mod p, and let $\alpha_3 = p\alpha_1, \ \alpha_4 = p\alpha_2$.

12. Take, for example: 1, 3, 5, 7, 2, 6, 10, 14.

13. In $\mathbb{Q}_5$ we have ± 1 and the two square roots of -1 found in Exercise 9. To prove the general fact, use Hensel's lemma for each $a_0 = 0, 1, \ldots, p - 1$ with $F(x) = x^p - x$.

14. See Herstein's *Topics in Algebra*, p. 160 (where it's proved for polynomials with integer coefficients; but the proof is the same with p-adic integer coefficients).

15. If there were such a pth root, then the polynomial $(x^p - 1)/(x - 1)$ would have a linear factor. But substituting $y = x + 1$ leads to an Eisenstein polynomial, which is irreducible by Exercise 14. To give the second proof, notice that $(1 + p^r x')^p = 1 + p^{r+1} x' + $ (terms divisible by p^{2r+1}), and this cannot equal 1.

16. Note that $1/(1 - p) - (1 + p + p^2 + \cdots + p^N) = p^{N+1}/(1 - p)$. The other two series converge to $1/(1 + p), \ (p^2 - p + 1)/(1 - p^2)$.

17. (a) More generally, in place of p^i one can take any sequence $p_i \in \mathbb{Z}_p$ such that $\text{ord}_p(p_i) = i$. Namely, show by induction on n that the map
$$\left\{ \begin{array}{l} \text{all sums } a_0 p_0 + \cdots + a_{n-1} p_{n-1} \\ \text{with varying digits } a_i \end{array} \right\} \to \{0, 1, 2, \ldots, p^n - 1\}$$
obtained by reducing modulo p^n, is one-to-one.
(b) We have
$$-(p - 1) \sum_{i < n, \, i \, \text{odd}} p^i \le a_0 + \cdots + a_{n-1}(-p)^{n-1} \le (p - 1) \sum_{i < n, \, i \, \text{even}} p^i;$$
since there are $1 + (p - 1) \sum_{i < n} p^i = p^n$ integers in this interval, by part (a) each such integer has exactly one representation in the form $a_0 + \cdots + a_{n-1}(-p)^{n-1}$.

18. Use Hensel's lemma with $F(x) = x^n - \alpha$ (or $\alpha x^{-n} - 1$ if $n < 0$) and $a_0 = 1$. $1 + p$ has no pth root. Next, if $\alpha = 1 + a_2 p^2 + \cdots$, then to find a pth root let $a_0 = 1 + a_2 p$ and apply Exercise 6 with $M = 1, \ F(x) = x^p - \alpha$.

19. Use induction on M to prove the congruence: if $\alpha^{p^{M-1}} = \alpha^{p^{M-2}} + p^{M-1}\beta$ for some β (this is the induction assumption), then raising both sides to the pth power gives the desired congruence. Then show that the limit as $M \to \infty$ has the properties (1) its pth power is itself, and (2) it's congruent to α mod p.

20. Use the same idea as in Exercise 19 of §2.

21. Look for $X = A_0 + pA_1 + p^2A_2 + \cdots$, where the A_i are $r \times r$ matrices whose entries are p-adic digits $0, 1, 2, \ldots, p - 1$. Let $X_n = A_0 + \cdots + p^nA_n$. We want X_n to satisfy $X^2 - AX + B \equiv 0 \mod p^{n+1}$ (where we use the congruence notation for matrices to denote entry-by-entry congruence). Use induction on n. When $n = 0$ we obtain the congruence by choosing $A_0 \equiv A \mod p$. The induction step is:

$$(X_{n-1} + p^nA_n)^2 - A(X_{n-1} + p^nA_n) + B$$
$$\equiv (X_{n-1}^2 - AX_{n-1} + B) + p^n(A_nX_{n-1} + X_{n-1}A_n - AA_n)$$
$$\equiv (X_{n-1}^2 - AX_{n-1} + B) + p^nA_nA \pmod{p^{n+1}},$$

because $X_{n-1} \equiv A \pmod{p}$. Choose A_n modulo p to be

$$-(X_{n-1}^2 - AX_{n-1} + B)p^{-n}A^{-1}.$$

Notice that this argument falls through for higher degree polynomials because of the noncommutativity of matrix multiplication.

Chapter II §2

1. Expand $1/(1 - q^{-s})$ in a geometric series; then multiply out, and use the fact that every positive integer n can be written (uniquely) as $q_1^{\alpha_1} \cdots q_r^{\alpha_r}$.

4. Define $f(t) = \frac{1}{2}t + t/(e^t - 1)$ and show that $f(t)$ is an even function, i.e., $f(t) = f(-t)$.

5. For large k, $\zeta(2k)$ is near 1. Answer: $4\sqrt{\pi k}(k/\pi e)^{2k}$.

6. (i) modulo 5^5 we have $(1 + 2 \cdot 5)^{1/(625 + 1 - 25)} \equiv (1 + 2 \cdot 5)^{1 + 5^2} = 11 \cdot (1 + 2 \cdot 5)^{5^2}$
$\equiv 1 + 2 \cdot 5 + 0 \cdot 5^2 + 2 \cdot 5^3 + 2 \cdot 5^4$;
(ii) modulo 3^5 we have $(1 + 3^2)^{-1/2} \equiv (1 + 3^2)^{1 + 3 + 3^2} \equiv 1 + 0 \cdot 3 + 1 \cdot 3^2 + 1 \cdot 3^3 + 1 \cdot 3^4$;
(iii) $1 + 5 \cdot 7 + 3 \cdot 7^2 + 2 \cdot 7^3 + 2 \cdot 7^4 + \cdots$.

7. Note that $p^N \equiv 1$ modulo $p - 1$ for any N, so that, if you first approximate a given p-adic integer by the nonnegative integer obtained from the first N places in its p-adic expansion, you can then add a suitable multiple of p^N to get a positive integer congruent to s_0 modulo $p - 1$ which is an equally good approximation.

9. $L_\chi(1) = \pi/4$; $L_\chi(s) = \prod 1/(1 - \chi(q)/q^s)$.

Chapter II §3

2. An example is the complement of a point.

5. It suffices to prove this for $U = a + (p^n)$, since any U is a disjoint union of such sets. Let a' be the least nonnegative residue mod p^n of αa; then since $|\alpha|_p = 1$, it follows that $\alpha U = a' + (p^n)$, and so both U and αU have the same Haar measure p^{-n}.

6. (1) the first digit in α; (2) $(p - 1)/2$; (3) $\Sigma_{a=0}^{p-1} a(a/p - 1/2) = (p^2 - 3p + 2)/12$.

Chapter II §5

1. For the first assertion, compare coefficients of t^k in the identity: $te^{tx}/(e^t - 1) = (\sum B_k t^k/k!)(\sum (tx)^j/j!)$. To prove the second assertion, take $\int_0^1 \cdots dx$ of both sides of the identity $\sum B_k(x)t^k/k! = te^{tx}/(e^t - 1)$ and compare coefficients of t^k. To get the third assertion, apply $(1/t)(d/dx)$ to both sides of the same identity.

2. *Claim:* $\mu(U) = 0$ for any U if μ has this property. Since U is a union of sets of the form $a + (p^N)$ for N arbitrarily large, we have for such N: $|\mu(U)|_p \leq \max_a|\mu(a + (p^N))|_p$. Now let $N \to \infty$.

3. $B_k; \; p^{k-1}B_k; \; (1 - p^{k-1})B_k$.

5. $(1 - \alpha^{-k})B_k; \; (1 - \alpha^{-k})(1 - p^{k-1})B_k; \; \sum_{i=0}^n a_i(1 - \alpha^{-i-1})(1 - p^i)B_{i+1}/(i + 1)$.

7. Use the corollary at the end of §5 with $g(x) = 1/(\text{the first digit in } x)$, so that $f(x) \equiv g(x) \bmod p$, and $g(x)$ is locally constant.

9.
$$\lim_{\substack{N \to \infty \\ N = 2M+1}} \sum_{\substack{0 \leq a < p^N \\ a = a_0 + a_2 p^2 + \cdots + a_{2M}p^{2M}}} a/p^{M+1}$$

$$= \lim_{M \to \infty} \frac{1}{p^{M+1}} \left(p^M \sum_{a_0=0}^{p-1} a_0 + p^M \sum_{a_2=0}^{p-1} a_2 p^2 + \cdots + p^M \sum_{a_{2M}=0}^{p-1} a_{2M}p^{2M} \right)$$

$$= \frac{p-1}{2}(1 + p^2 + p^4 + \cdots) = -\frac{1}{2(p+1)}.$$

CHAPTER II §7

2. (i) Use the Kummer congruence $(1 - 5^{2-1})(-B_2/2) \equiv (1 - 5^{102-1})(-B_{102}/102)$ modulo 5^3 to obtain $1/3 \equiv -B_{102}/102 \pmod{5^3}$, and hence $B_{102} = 1 + 3 \cdot 5 + 3 \cdot 5^2 + \cdots$.

   (ii) From the congruence $(1 - 7^{2-1})(-B_2/2) \equiv (1 - 7^{296-1})(-B_{296}/296) \pmod{7^3}$, obtain $B_{296} = 6 + 6 \cdot 7 + 3 \cdot 7^2 + \cdots$.

   (iii) Use $(1 - 7^{4-1})(-B_4/4) \equiv (1 - 7^{592-1})(-B_{592}/592) \pmod{7^3}$ to obtain $B_{592} = 3 + 4 \cdot 7 + 3 \cdot 7^2 + \cdots$.

3. Recall that a rational number belongs to $\mathbb{Z}$ if and only if for each p it is in $\mathbb{Z}_p$. Then use parts (1) and (3) of Theorem 7.

6. Let $\alpha = 1 + p^2 = 5$, and let $g(x) = (a_0 + 2a_1)^{-1}$, where a_0, a_1 are the first two 2-adic digits in x. Then follow the proof given for odd p. In the case $p = 2$, the Clausen–von Staudt theorem says that every nonzero Bernoulli number starting with B_1 has exactly one power of 2 in the denominator.

CHAPTER III §1

1. All roots of $X^{p^{f'}-1} - 1$ are also roots of $X^{p^f-1} - 1$ if and only if $X^{p^{f'}-1} - 1$ divides $X^{p^f-1} - 1$; this is true if and only if $p^{f'} - 1$ divides $p^f - 1$, which, in turn, is equivalent to f' dividing f.

2. Here is a table of all possible generators of $\mathbb{F}_p^\times$:

p	2	3	5	7	11	13
possible a	1	2	2, 3	3, 5	2, 6, 7, 8	2, 6, 7, 11

3. $(1 + j)^{\text{any odd power}}$ is a generator.

4. Adjoin a root j of $X^2 + X + 1 = 0$ and $X^3 + X + 1 = 0$, respectively. For example, in $\mathbb{F}_8$ we multiply as follows:

$$(a + bj + cj^2)(d + ej + fj^2) = (ad + bf + ce) + (ae + bd + bf + ce + cf)j$$
$$+ (af + be + cd + cf)j^2.$$

Finally, note that when $q - 1$ is prime, any element (not 1) of $\mathbb{F}_q^\times$ is a generator.

5. Clearly $\mathbb{F}_p(a) = \mathbb{F}_q$.

6. If any two of the σ_i were the same, you would get a polynomial of degree less than q having q roots.

7. If $P(X)$ factored over $\mathbb{F}_p$, say, $P(X) = P_1(X)P_2(X)$ with $\deg P_1 = d < p$, then the coefficient of X^{d-1} in $P_1(X)$, which is minus the sum of d of the roots $a + i$, would be in $\mathbb{F}_p$. But then $da \in \mathbb{F}_p$, and so $a \in \mathbb{F}_p$. But all elements of $\mathbb{F}_p$ are roots of $X^p - X$, and so cannot be roots of $P(X)$.

8. If $p = 2$, then $-1 = 1$ and it's trivial; otherwise $\mathbb{F}_q$ contains a square root of -1 if and only if 1 has a primitive fourth root, i.e., if and only if 4 divides $q - 1$.

9. Assume the contrary, and use the same approach as in Exercise 19 of §I.2 and Exercise 20 of §I.5.

10. First show, without using limits, that the formal derivative of a polynomial over any field obeys the product rule. This can be done quickly by using linearity of the derivative to reduce to the case of a product of two powers of X.

CHAPTER III §2

A good reference for the ideas in these exercises (especially Exercises 2, 6, 7, 8) is Chapter IV of Simmons' textbook (see Bibliography).

3. $v_2 \cdot v_2 = pv_1$, but $\|v_2\|_{\sup} \cdot \|v_2\|_{\sup} = 1$, $\|pv_1\|_{\sup} = |p|_p = 1/p$.

4. Let $\mathbb{F}_q = A/M$ be the residue field of K, where $q = p^f$; it is an extension of degree f of the residue field $\mathbb{F}_p$ of $\mathbb{Q}_p$. In the proof of the last proposition we saw that $f \leq n = [K:\mathbb{Q}_p]$. (In the next section we'll see that f divides n; $e = n/f$ is called the ramification index.) First suppose that K is *unramified*, i.e., $f = n$. If we let $\bar{x}$ denote the image of x under the residue map $A \to \mathbb{F}_q$, we see that we can choose a basis $\{v_1, \ldots, v_n\}$ of K over $\mathbb{Q}_p$ such that $\{\bar{v}_1, \ldots, \bar{v}_n\}$ is a basis of $\mathbb{F}_q$ over $\mathbb{F}_p$. One now checks that the sup-norm with respect to such a basis has the desired multiplicative property. Namely, first prove that $x \in K$ we have: $\|x\|_{\sup} \leq 1$ if and only if $x \in A$; $\|x\|_{\sup} < 1$ if and only if $x \in M$. Then to show that $\|xy\|_{\sup} = \|x\|_{\sup} \cdot \|y\|_{\sup}$, reduce to the case when $\|x\|_{\sup} = \|y\|_{\sup} = 1$, i.e., $x, y \in A - M$. But then $xy \in A - M$. Conversely, if K is ramified, the sup-norm is *never* a field norm. Namely, one can show that in a field norm K has elements with norm a fractional power of p.

5. Any element $x \in \mathbb{Z}_p$ can be written in the form $x = p^n u$, where u is a unit.

CHAPTER III §4

1. The values of $|\ |_p$ on Ω are the same as on $\overline{\mathbb{Q}}_p$, since any element can be approximated by an element of $\overline{\mathbb{Q}}_p$. To show that, for example, the unit ball in $\overline{\mathbb{Q}}_p$ is not sequentially compact, take any sequence of distinct roots of unity of order prime to p and show that it has no convergent subsequence.

2. Let $r_0 = |b - a|_p$. You get the empty set unless r is a sum $r_1 + r_2$ of two rational powers of p (or zero). Then consider cases depending on the relative size of r_0, r_1, r_2. For example, if $r_0 = r_1 > r_2$, then you get the two disjoint circles of radius r_2 about a and b. The "hyperbola" has exactly the same possibilities; now $r = r_1 - r_2$ must be a difference of two rational powers of p.

3. Let $C_1 = \max(1, C_0)$. Suppose β is a root with $|\beta|_p > C_1$. Then $\beta = -b_{n-1} - b_{n-2}/\beta - \cdots - b_0/\beta^{n-1}$, and $|\beta|_p \le \max(|b_{n-i-1}/\beta^i|_p) \le \max(|b_i|_p) = C_0$, a contradiction.

4. Set $\delta = \min|\alpha - \alpha_i|_p$, where the minimum is over all roots $\alpha_i \ne \alpha$ of f. Use the last proposition in this section with the roles of δ and ε reversed to find a root β of g such that $|\alpha - \beta|_p < \delta$. By Krasner's lemma, $K(\alpha) \subset K(\beta)$. Since f is irreducible, $[K(\alpha):K] = \deg f = \deg g \ge [K(\beta):K]$, and hence $K(\alpha) = K(\beta)$. As a counterexample when f is no longer irreducible, take, say, $K = \mathbb{Q}_p$, $f(X) = X^2$, $\alpha = 0$, $g(X) = X^2 - p^{2N+1}$ for large N.

5. Let α be a primitive element, i.e., $K = \mathbb{Q}_p(\alpha)$, and let $f(X) \in \mathbb{Q}_p[X]$ be its monic irreducible polynomial. Choose ε as in Exercise 4, and find $g(X) \in \mathbb{Q}[X]$ such that $|f - g|_p < \varepsilon$. (For example, take the coefficients of g to be partial sums of the p-adic expansions of the corresponding coefficients of f.) Then g has a root β such that $K = \mathbb{Q}_p(\beta) \supset \mathbb{Q}(\beta)$, and it's easy to see that $F = \mathbb{Q}(\beta)$ is dense in K and has degree $n = [K:\mathbb{Q}_p]$ over $\mathbb{Q}$.

6. Set $\alpha = \sqrt{-1}$, $\beta = \sqrt{-a}$ (with any fixed choice of square roots). We can apply Krasner's lemma if either $|\beta - \alpha|_p$ or $|\beta - (-\alpha)|_p$ is less than $|\alpha - (-\alpha)|_p$, which equals 1 if $p \ne 2$ and $1/2$ if $p = 2$. Since

$$|a - 1|_p = |-a - (-1)|_p = |\beta - \alpha|_p |\beta + \alpha|_p,$$

this holds if $|a - 1|_p < 1$ for $p \ne 2$ and $< 1/4$ for $p = 2$. To do the next part, set $\alpha = \sqrt{p}$, $\beta = \sqrt{a}$. Then it suffices to have either $|\beta - \alpha|_p$ or $|\beta - (-\alpha)|_p$ less than $|2\sqrt{p}|_p$. Since $|a - p|_p = |\beta - \alpha|_p \cdot |\beta + \alpha|_p$, this holds if $|a - p|_p < |4p|_p$. So choose $\varepsilon = 1/p$ if $p \ne 2$ and $= 1/8$ if $p = 2$.

7. First note that a satisfies the monic irreducible polynomial $(X^{p^n} - 1)/(X^{p^{n-1}} - 1)$. (For the case $n = 1$, see Exercise 15 of §I.5.) Now let $\beta = (-p)^{1/(p-1)}$, i.e., β is a fixed root of $X^{p-1} + p = 0$. Let $\alpha_1 = a - 1, \alpha_2 = a^2 - 1, \ldots, \alpha_{p-1} = a^{p-1} - 1$ be the conjugates of $a - 1$. Check that $|\alpha_i - \alpha_j|_p = p^{-1/(p-1)}$ for any $i \ne j$. By Krasner's lemma, it suffices to show that $|\beta - \alpha_i|_p < p^{-1/(p-1)}$ for some i. If this were not the case, we would have $p^{-1} \le \prod_{i=1}^{p-1} |\beta - \alpha_i|_p = |((\beta + 1)^p - 1)/\beta|_p$, since $\prod(X - \alpha_i) = ((X + 1)^p - 1)/X$. Now use the relation $\beta^{p-1} + p = 0$ to obtain: $((\beta + 1)^p - 1)/\beta = \beta \cdot \sum_{i=2}^{p-1} \binom{p}{i}\beta^{i-2}$. But the p-adic norm of this is bounded by $|p\beta|_p < p^{-1}$. To prove the last assertion in the exercise, suppose that a is a primitive mth root of 1, with m not a power of p, and $|a - 1|_p < 1$. Then we would have $|a^i - 1|_p < 1$ for any i. Let $l \ne p$ be a prime factor of m, and let $b = a^{m/l}$, which is a primitive lth root of unity.

Then $\beta = b - 1$ satisfies: $|\beta|_p < 1$, and at the same time $0 = ((\beta + 1)^l - 1)/\beta = \sum_{i=2}^l \binom{l}{i}\beta^{i-1} + l$. But then $|l|_p = |\beta(-\sum_{i=2}^l \binom{l}{i}\beta^{i-2})|_p < 1$, a contradiction.

8. Let π be an element of K with $\mathrm{ord}_p \pi = 1/e$, where e is the ramification index of K. Then $\{\pi^i\}_{i=0,1,\cdots,m-1}$ are in distinct multiplicative cosets modulo $(K^\times)^m$, and any element of $K^\times$ can be written uniquely as $\pi^{i+mj}u$ for some $0 \le i < m$, $j \in \mathbb{Z}$, $u \in K$ with $|u|_p = 1$. Now show that u is an mth power. Namely, since its image in the residue field $\mathbb{F}_q$ is an mth power, we can find u_0 such that $\alpha = u/u_0^m - 1$ satisfies $|\alpha|_p < 1$. Finally, write the p-adic expansion $1/m = a_0 + a_1 p + a_2 p^2 + \cdots$, and obtain

$$u = u_0^m(1 + \alpha) = (u_0(1 + \alpha)^{a_0 + a_1 p + a_2 p^2 + \cdots})^m.$$

10. Otherwise, K would have residue degree $f > 1$, since it would have more than $p - 1$ roots of unity of order prime to p, and any two such roots have distinct residues in $\mathbb{F}_{p^f}^\times$.

11. All have the cardinality of the continuum.

12. Let $y_1, \ldots, y_f$ be elements in K such that $|y_i|_p = 1$ and the images of the y_i in the residue field form a basis of the residue field over $\mathbb{F}_p$. Show that $y_i \pi^j$, $1 \le i \le f$, $0 \le j \le e - 1$, form a basis of K over $\mathbb{Q}_p$, where $\mathrm{ord}_p \pi = 1/e$.

13. If β satisfies the Eisenstein polynomial $X^e + a_{e-1}X^{e-1} + \cdots + a_0$, set $\alpha = -a_0$. Then $\alpha \in \mathbb{Z}_p$, $\mathrm{ord}_p \alpha = 1$, and $\beta^e - \alpha = \beta^e + a_0 = -a_{e-1}\beta^{e-1} - \cdots - a_1\beta$ has p-adic norm less than $1/p$.

14. Follow the proof of Theorem 3 of Ch. I but working over the field K, with β playing the role of p (recall that $\mathrm{ord}_p \beta = 1/e$). Note that $\mathrm{ord}_p(\beta^e - \alpha) \ge 1 + 1/e$. Look for $\beta + a_2\beta^2$ with $a_2 \in \{0, 1, \ldots, p - 1\}$ such that $\mathrm{ord}_p((\beta + a_2\beta^2)^e - \alpha) \ge 1 + 2/e$, and so on. An alternate method is to note that $|\alpha/\beta^e - 1|_p < 1$, then write the p-adic expansion for $1/e \in \mathbb{Z}_p$, and compute $\beta(\alpha/\beta^e)^{1/e} \in K$; this will be an eth root of α. Finally, we have $K = \mathbb{Q}_p(\beta)$ because β has degree e.

15. Let $V \subset \mathbb{Z}_p[X]$ be the set of monic polynomials of degree n. For $f, g \in V$ define $|f - g|_p$ by the sup-norm. Note that V looks like $\mathbb{Z}_p^n$ with the sup-norm, and it is compact. Let $S \subset V$ be the subset consisting of irreducible polynomials. Any such polynomial gives at most n different degree n extensions of $\mathbb{Q}_p$ in $\overline{\mathbb{Q}}_p$. For fixed $f \in S$, the last two propositions in §3 show that there exists $\delta > 0$ such that any $g \in S$ with $|f - g|_p < \delta$ gives precisely the same set of degree n extensions as does f. By compactness, the set S has a finite covering of subsets in each of which the polynomials give the same extensions. So there are only finitely many extensions of degree n.

16–17. See the article "Algebraic p-adic expansions" by David Lampert, to appear in the *Journal of Number Theory*.

CHAPTER IV §1

1. (i) $D(p^{1/(p-1)-})$; (ii) $D(\infty)$ = all of Ω; (iii) $D(p^-)$; (iv) $D(1)$; (v) $D(1)$; (vi) $D(1^-)$; (vii) $D(1^-)$.

3. A counterexample for the last question: let $f_j = 1 + pX$ if j is a power of 2, and let $f_j = 1$ otherwise. Then $f(X) = \prod_{k=0}^{\infty} (1 + pX^{2^k}) = \sum_{j=0}^{\infty} p^{S_j} X^j$, where S_j denotes the sum of the binary digits in j; this does not converge on $D(1)$.

4. Let $d(x, \mathbb{Z}_p) \underset{\text{def}}{=} \min\{|x - y|_p \,|\, y \in \mathbb{Z}_p\}$, i.e., the "distance" from x to $\mathbb{Z}_p$. First prove that if $d(x, \mathbb{Z}_p) \geq 1$, then the series converges, in fact, it converges under a much weaker condition on $|a_n|_p$. Now suppose that $d(x, \mathbb{Z}_p) = r < 1$. Choose M so that $p^{-(M+1)} \leq r < p^{-M}$. Then for $n = p^N$ with $N > M$, show that: $p^N - p^{N-1}$ of the factors in the denominator of the nth term are of norm 1, $p^{N-1} - p^{N-2}$ of the factors are of norm $1/p$, $p^{N-2} - p^{N-3}$ of the factors are of norm $1/p^2$, and so on, and finally $p^{N-M} - p^{N-M-1}$ of the factors are of norm $1/p^M$, and the remaining p^{N-M-1} factors are of norm r. Use this to give a lower bound for the ord_p of the nth term, namely,

$$\text{ord}_p \, a_n + \text{ord}_p \, n! - (p^{N-1} - p^{N-2}) - 2(p^{N-2} - p^{N-3}) - \cdots$$
$$- M(p^{N-M} - p^{N-M-1}) - (M+1)p^{N-M-1}$$
$$= \text{ord}_p \, a_n + \text{ord}_p \, n! - n(p^{-1} + p^{-2} + \cdots + p^{-M-1})$$
$$= \text{ord}_p \, a_n + (n/p^{M+1} - 1)/(p - 1).$$

The more general case when n is not a power of p involves the same sort of estimate. In all cases, one finds that ord_p of the nth term approaches infinity.

6. For $p > 2$, write the congruence in the form $((1 + 1)^p - 2)/p \equiv 0 \pmod{p}$, use the binomial expansion for $(1 + 1)^p$ to get $\sum_{j=1}^{p-1} (p - 1)(p - 2) \cdots (p - j + 1)/j!$ on the left, and consider each term modulo p.

7. Let $a = \log_2(1 - 2) = -\lim_{n \to \infty} \sum_{i=1}^{n} 2^i/i$. Now $2a = \log_2((1 - 2)^2) = \log_2 1 = 0$; hence, $a = 0$. Thus, $\text{ord}_2 \sum_{i=1}^{n} 2^i/i = \text{ord}_2 \sum_{i=n+1}^{\infty} 2^i/i \geq \min_{i \geq n+1} (i - \text{ord}_2 i)$. For example, for $n = 2^m$ this minimum is $n + 1$.

8–9. Non-theorem 1 is being used.

10. (a) The real series for the square root converges to the positive square root $(m + p)/m$; the p-adic series converges to the square root which is congruent to 1 modulo p. Here they're both the same.
(b) Here the positive square root is $(p + 1)/2n = (p + 1)/(p - 1)$, but this is the *negative* of the square root which is $\equiv 1 \bmod p$.

11. (d)–(e) Consider the following cases separately:
Case (i). $a - b$ is divisible by an odd prime p. Note that, since a and b are positive, relatively prime, and not both 1, it follows that $a + b \geq 3$ is divisible either by 4 or by an odd prime $p_1 \neq p$. Then $(1 + \alpha)^{1/2}$ converges to a/b p-adically and converges to $-a/b$ either 2-adically (if 4 divides $a + b$) or p_1-adically (if p_1 divides $a + b$).
Case (ii). $a - b = \pm 2^r, r \geq 2$. Note that in this case there must be an odd prime p dividing $a + b$. Then $(1 + \alpha)^{1/2}$ converges to a/b 2-adically and to $-a/b$ p-adically.
Case (iii). $a - b = \pm 2$, so that $\alpha = (a/b)^2 - 1 = 4(\pm b + 1)/b^2$. Note that here a and b are both odd. Then $(1 + \alpha)^{1/2}$ converges to $-a/b$ 2-adically and to a/b in the reals, provided that $-1 < \alpha < 1$. The latter inequality holds unless $b = 1, \alpha = 8$ or $b = 3, \alpha = 16/9$.
Case (iv). $a - b = \pm 1$, so that $\alpha = (\pm 2b + 1)/b^2$. Then $(1 + \alpha)^{1/2}$ converges to $-a/b$ p-adically for p a prime dividing $\pm 2b + 1$, and it converges to a/b in the reals unless $b = 1, \alpha = 3$ or $b = 2, \alpha = 5/4$.

12. The series $(x \, d/dx)^k 1/(1 - x) = \sum n^k x^n$ represents a polynomial in $\mathbb{Z}[X]$ divided by $(1 - X)^{k+1}$ in its disc of convergence, in both the real and p-adic situations. In particular, for $x = p$ you get an integer divided by $(p - 1)^{k+1}$.

13. The left side is $-\log_3(1 + \frac{9}{16}) = -\log_3(\frac{25}{16})$; the right side is

$$-2 \log_3(1 - \tfrac{9}{4}) = -\log_3((-\tfrac{5}{4})^2) = -\log_3(\tfrac{25}{16}).$$

14. As an example where the exact regions of convergence differ, take $f(X) = \sum X^{p^n}$. Then $f(X)$ converges on $D(1^-)$ and $f'(X)$ converges on $D(1)$.

15. (a) For example, $\sum_{i=1}^{\infty} i!/p_i^i$, where p_i denotes the ith prime. (b) I don't know of an example, or of a proof that it's impossible.

17. For each rational number $r = a/b \in \mathbb{Q}$, make a choice of $p^r \in \Omega$, i.e., choose a root of $x^b - p^a = 0$ and denote it p^r. Now take, for example, $f(x) = p^{(\mathrm{ord}_p x)^2}$.

18. No.

20. If you want each coefficient to j places, choose N so that $p^N > Mp^{j-1}$; write $a/b \in \mathbb{Z}_p$ modulo p^N in the form $a_0 + a_1 p + \cdots + a_{N-1} p^{N-1}$; and compute the coefficients of $(1 + X)^{a_0 + a_1 p + \cdots + a_{N-1} p^{N-1}}$ modulo p^j.

21. (6) First prove that the convergence assumptions allow you to rearrange terms. Reduce everything to proving than an element in $\mathbb{R}[[X]]$ which vanishes for values of the variables in $[-\varepsilon', \varepsilon']$ must be the zero power series. Prove this fact by induction on n.

Chapter IV §2

1. $6 \cdot 7 + 2 \cdot 7^2 + 5 \cdot 7^3$; $2^4 + 2^5 + 2^6 + 2^8 + 2^9 + 2^{10} + 2^{11}$.

2. By removing roots of unity, show that the image of $\mathbb{Z}_p$ is the same as the image of $1 + p\mathbb{Z}_p$ for $p > 2$ and $1 + 4\mathbb{Z}_2$ for $p = 2$.

3. $p^2 | \log_p a \Leftrightarrow p^2 | (p - 1)\log_p a = \log_p(a^{p-1})$, and the latter is congruent to $a^{p-1} - 1 \bmod p^2$, since in general $\log_p(1 + x) \equiv x \bmod p^2$ for $x \in p\mathbb{Z}_p$.

4. $\log_p x$ (no surprise!).

5. Let $c = f(p)$, and show that $f(x) - c \, \mathrm{ord}_p x$ satisfies all three properties which characterize $\log_p x$.

8. $|f(1 + p^N) - f(1)|_p = |-1 - 1|_p = 1$.

9. $j^2 - 1 = (j + 1)(j - 1)$, and for $p > 2$ exactly one of the two factors is divisible by p, and hence by p^N. If $p = 2$, then you have $j \equiv \pm 1 \bmod 2^{N-1}$.

10. Approximate $1/2$ by $(p^N + 1)/2$, and compare $(\prod_{j < (p^N + 1)/2, \, p \nmid j} j)^2$ with $\prod_{j < p^N, \, p \nmid j} j$, which we proved is $\equiv -1 \pmod{p^N}$.

12. In the first equality both sides are $1 + 3 \cdot 5 + 2 \cdot 5^2 + 3 \cdot 5^3 + \cdots$, and in the second equality both sides are $1 + 6 \cdot 7 + 5 \cdot 7^2 + 4 \cdot 7^3 + \cdots$.

13. On the right side of property (3), we have $s_0 = p - r$, $s_1 = (p - 1 - r)/(1 - p)$. Note that the expression is congruent to m^{1-s_0} mod p; then it remains to verify that the $(p - 1)$th power is 1. Use: $(p - 1)s_1 = 1 - p + r = 1 - s_0$.

14. In all cases show that the image of the function is in the open unit disc about 1.

15. $1 + X + X^2 + \frac{2}{3}X^3 + \frac{2}{3}X^4$; $1 + X + \frac{1}{2}X^2 + \frac{1}{2}X^3 + \frac{3}{8}X^4$.

16. The coefficient of X^p is $((p - 1)! + 1)/p!$; Wilson's theorem.

17. $E_p(X^p)/E_p(X)^p = e^{-pX} \in 1 + pX\mathbb{Z}_p[[X]]$.

18. $f(X^p)/f(X)^p = \exp(\sum_{i=0}^{\infty}(b_{i-1} - pb_i)X^{p^i})$. If $c_i \underset{\text{def}}{=} b_{i-1} - pb_i \in p\mathbb{Z}_p$ for all i, then since $e^{cX} \in 1 + pX\mathbb{Z}_p[[X]]$ whenever $c \in p\mathbb{Z}_p$, it follows that

$$\prod e^{c_i X^{p^i}} \in 1 + pX\mathbb{Z}_p[[X]].$$

Conversely, suppose c_{i_0} is the first c_i not in $p\mathbb{Z}_p$; then the coefficient of $X^{p^{i_0}}$ in $\prod e^{c_i X^{p^i}}$ is congruent mod p to $c_{i_0} \not\equiv 0$ mod p, and by Dwork's lemma $f(X) \notin 1 + X\mathbb{Z}_p[[X]]$.

CHAPTER IV §4

1. (i) Join $(0, 0)$ to $(1, 0)$ and $(1, 0)$ to $(2, 1)$.
    (ii) Join $(0, 0)$ to $(3, -2)$.
    (iii) Join $(0, 0)$ to $(2, 0)$, $(2, 0)$ to $(4, 1)$, and $(4, 1)$ to $(6, 3)$.
    (iv) Join $(0, 0)$ to $(p - 2, 0)$ and $(p - 2, 0)$ to $(p - 1, 1)$.
    (v) Join $(0, 0)$ to $(1, 0)$, $(1, 0)$ to $(2, 1)$, and $(2, 1)$ to $(3, 4)$.
    (vi) Join $(0, 0)$ to $(p^2 - p, 0)$, $(p^2 - p, 0)$ to $(p^2 - 1, p - 1)$, and $(p^2 - 1, p - 1)$ to $(p^2, p + 1)$.

2. (a) Any root α of $f(X)$ that satisfied a polynomial of lower degree d would have $\text{ord}_p \alpha$ equal to a fraction with denominator at most d; but by Lemma 4, $\text{ord}_p \alpha = -m/n$.
    (b) If $f(X)$ is an Eisenstein polynomial, then $a_n^{-1}X^n f(1/X) = 1 + a_{n-1}/a_n X + \cdots + a_0/a_n X^n$ has for its Newton polygon the line joining $(0, 0)$ to $(n, 1)$.
    (c) Counterexample: $1 + pX + ap^2X^2$, where $a \in \mathbb{Z}_p^\times$ is chosen so that $1 - 4a$ is not a square in $\mathbb{Z}_p$.

3. All slopes are between 0 and 1, and for each segment of slope λ there's a segment (of the same horizontal length) with slope $1 - \lambda$. The number of possible Newton polygons of this type is: 2 for $n = 1$; 3 for $n = 2$; 5 for $n = 3$; 8 for $n = 4$.

4. (i) Join $(p^j - 1, -j)$ to $(p^{j+1} - 1, -(j + 1))$ for $j = 0, 1, 2, \ldots$.
    (ii) The horizontal line from $(0, 0)$.
    (iii) Join $(p^j - 1, 1 + p + \cdots + p^{j-1} - j)$ to
    $(p^{j+1} - 1, 1 + p + \cdots + p^{j-1} + p^j - (j + 1))$ for $j = 0, 1, 2, \ldots$
    (iv) One infinite straight line from $(0, 0)$ with slope $-1/(p - 1)$.
    (v) The segment joining $(0, 0)$ to $(2, 1)$ and the infinite line from $(2, 1)$ with slope 1.
    (vi) The infinite line from $(0, 0)$ with slope 1.
    (vii) Join $(j, 1 + 2 + \cdots + (j - 1))$ to $(j + 1, 1 + 2 + \cdots + (j - 1) + j)$ for $j = 0, 1, 2, \ldots$.

(viii) Starting with a segment from $(0, 0)$ to $(2, 2)$, there are infinitely many segments with slope increasing toward $\sqrt{2}$; the details of these segments depend upon rational approximations to $\sqrt{2}$.

5. The Newton polygon of $1 + \sum_{i=1}^{\infty} p^{1+[i\sqrt{2}]}X^i$ is the infinite line from $(0, 0)$ with slope $\sqrt{2}$.

6. For example, $\sum_{i=0}^{\infty} p^i X^{p^i-1}$ converges on $D(1)$, its Newton polygon is the horizontal line from $(0, 0)$, and it has no zero in $D(1)$.

8. Reduce to the case $\lambda = 0$ by replacing $f(X)$ by $f(p^{-\lambda}X)$, where p^λ is a choice of fractional power of p in Ω. Then multiply by a scalar to reduce to the case when

$$\min \operatorname{ord}_p a_i = 0.$$

For $x \in D(1)$ clearly $|f(x)|_p \le 1$. To obtain an x for which $|f(x)|_p = 1$, it suffices to reduce modulo the maximal ideal of Ω, i.e., to consider the series $\bar{f}(X) \in \bar{\mathbb{F}}_p[X]$. (This has only finitely many terms, because $\operatorname{ord}_p a_i \to \infty$.) Then choose any x such that $\bar{f}(\bar{x}) \ne 0$ in $\bar{\mathbb{F}}_p$.

9. Apply the Weierstrass Preparation Theorem to the series $f_1(X) = f(X)/a_n X^n \in 1 + X\Omega[[X]]$ which is obtained by dividing $f(X)$ by its leading term $a_n X^n$. Take $\lambda = 0$.

10. Reduce to the case $f(X) \in 1 + X\Omega[[X]]$ by factoring out the leading term, as in Exercise 9. Use the Weierstrass Preparation Theorem to write $h(X) = f(X)g(X)$. But $f(x) = 0$ implies that $h(x) = 0$, and $h(X)$ is a polynomial.

11. Use Exercises 9 and 10, and show that if E_p has one zero, then it must have infinitely many. To do this, let x be any pth root of a zero of E_p, and use the relation $E_p(X)^p = E_p(X^p)e^{pX}$.

12. Write $f(X)g(X) = h(X)$. If $f(X)$ has a coefficient a_i with $\operatorname{ord}_p a_i < 0$, then, by Lemma 4, $f(X)$ has a root α in $D(1^-)$. But then $h(\alpha) = 0$; however, f and h have no common roots. If $h(X)$ has a coefficient a_i with $\operatorname{ord}_p a_i < 0$, then it has a root α in $D(1^-)$. Since $g(\alpha) \ne 0$ it follows that $f(\alpha) = 0$, and we again have a contradiction.

CHAPTER V §1

1. $1/(1 - T)$; $1/(1 - q^n T)$.

2. $1/(1 - T)(1 - qT) \cdots (1 - q^n T)$.

3. There is a one-to-one correspondence between the points of $(n - 1)$-dimensional affine space and the points of H_f given by

$$(x_1, \ldots, x_{n-1}) \mapsto (x_1, \ldots, x_{n-1}, -g(x_1, \ldots, x_{n-1})).$$

4. Suppose, for example, that $r = 2$. Then show that $\# H_{(f_1, f_2)}(\mathbb{F}_{q^s}) = \# H_{f_1}(\mathbb{F}_{q^s}) + \# H_{f_2}(\mathbb{F}_{q^s}) - \# H_{f_1 f_2}(\mathbb{F}_{q^s})$, where $H_{f_1 f_2}$ is the hypersurface defined by the *product* of the two polynomials.

5. Write an n-dimensional projective hypersurface as a disjoint union of affine hypersurfaces (one in each dimension n, $n - 1$, $n - 2, \ldots$).

6. Let $Q(T) = \prod (1 - \alpha_i T)$, $P(T) = \prod (1 - \beta_i T)$; these are both in $\mathbb{Q}[T]$; then $\exp(\sum N_s T^s/s) = P(T)/Q(T)$. The converse is shown by reversing this procedure.

7. Comparing coefficients of T gives $a = N_1 - 1 - q$, where $N_1 = \#\tilde{E}(\mathbb{F}_q)$; by Exercise 6, $N_s = 1 + q^s - \alpha_1^s - \alpha_2^s$, where α_1, α_2 are given by $(1 + aT + qT^2) = (1 - \alpha_1 T)(1 - \alpha_2 T)$.

8. (i) For $q \equiv 2 \pmod 3$, every element of $\mathbb{F}_q$ has a unique cube root; then show that $N_1 = q + 1$, so that $a = 0$, and the zeta-function is $(1 + qT^2)/(1 - T)(1 - qT)$.
   (ii) For $q \equiv 3 \pmod 4$, -1 does not have a square root. Then for exactly one from each pair $x_1 = a, x_1 = -a$ we have a solution to $x_2 = \pm\sqrt{x_1^3 - x_1}$; this gives a one-to-one correspondence between $a \in \mathbb{F}_q$ and points (x_1, x_2) on the affine curve. Counting the point at infinity, we obtain $N = q + 1$, as in part (i), and the zeta-function is $(1 + qT^2)/(1 - T)(1 - qT)$. Next, for $q = 9 = 3^2$, we have $N_1 = $ (the N_2 when $q = 3$) $= 16$, $a = 6$, $Z(T) = (1 + 3T)^2/(1 - T)(1 - 9T)$. For $q = 5$ we have $Z(T) = (1 + 2T + 5T^2)/(1 - T)(1 - 5T)$, and for $q = 13$ we have $Z(T) = (1 - 6T + 13T^2)/(1 - T)(1 - 13T)$.

9. Suppose we have two rational functions $f(T)/g(T)$ and $u(T)/v(T)$ with numerator of degree m and denominator of degree n, where the first is $\exp(\sum_{s=1}^{\infty} N_s T^s/s)$ and the second is $\exp(\sum_{s=1}^{\infty} N_s' T^s/s)$, and suppose that $N_s = N_s'$ for $s = 1, 2, \ldots, m + n$. It suffices to show that then $f(T)/g(T) = u(T)/v(T)$, because that implies that $N_s = N_s'$ for all s. But $f(T)v(T) = g(T)u(T) \exp(\sum_{s=1}^{\infty}(N_s - N_s')T^s/s) = g(T)u(T) \times \exp(\sum_{s=m+n+1}^{\infty}(N_s - N_s')T^s/s)$, and the equality of polynomials comes from comparing coefficients of powers of T up to T^{m+n}.

10. Show that there are $(q - 1)q^2$ four-tuples with nonzero x_3 and $(q - 1)q$ with zero x_3; then the zeta-function is $(1 - qT)/(1 - q^3 T)$.

11. The zeta-function of the affine curve is listed first, followed by the zeta-function of its projective completion:
    (i) $(1 - T)/(1 - qT)^2$; $1/(1 - T)(1 - qT)^2$.
    (ii) $(1 - T)^3/(1 - qT)^3$; $1/(1 - qT)^3$.
    (iii) $(1 - T)/(1 - qT)$ (unless $p = 2$, in which case it is $1/(1 - qT)$);

    $$1/(1 - T)(1 - qT).$$

    (iv) $1/(1 - qT)$; $1/(1 - T)(1 - qT)$.
    (v) $(1 - T)/(1 - qT)$; $1/(1 - qT)$ (unless $p = 2$, in which case $1/(1 - qT)$; $1/(1 - T)(1 - qT)$).

12. $1/(1 - T)(1 - qT)(1 - q^2 T)^2(1 - q^3 T)(1 - q^4 T)$.

13. It suffices to show that for any prime p, the coefficients (which are *a priori* in $\mathbb{Q}$) are in $\mathbb{Z}_p$.

14. Write the numerator in the form $1 + a_1 T + a_2 T^2 + a_3 T^3 + p^2 T^4$. To show that $a_1 = a_2 = a_3 = 0$, i.e., that the zeta-function agrees with the zeta-function of the projective line through the T^3-term, it suffices to show that $\tilde{N}_s = p^s + 1$ for $s = 1, 2, 3$. But since $p^s \not\equiv 1 \pmod 5$ every element of $\mathbb{F}_{p^s}$ has a (unique) 5th root. (This is the same procedure as in Exercise 8(i).)

Answers and hints for the exercises

CHAPTER V §4

1. See the proof of the second proposition in §III.3.

2. Since $\Theta(T) = \sum a_j T^j$ with $\text{ord}_p a_j \geq j/(p-1)$, we have $\Theta(aX_1^{w_1} \cdots X_n^{w_n}) = \sum_{v=jw} g_v X^v$ with $\text{ord}_p g_v \geq |v|/(|w|(p-1))$. Set $M = 1/(|w|(p-1))$; then $\text{ord}_p g_v \geq M|v|$.

3. Use induction on the number of nonzero a_i's; in case of difficulty, see Lang's *Algebra*, p. 209.

4. Make a change of variables $x \mapsto x + x_0$, where $x_0 \in \mathbb{F}_q$ has nonzero trace.

5. For $u \neq 0$ make the change of variables $x \mapsto ux$ in Exercise 4.

Index

Index

Index

Graduate Texts in Mathematics

Graduate Texts in Mathematics